Crop Improvement for Sustainable Agriculture

Edited by
M. Brett Callaway and
Charles A. Francis

Crop Improvement for Sustainable Agriculture

University of Nebraska Press
Lincoln and London

Library of Congress Cataloging-in-Publication Data
Crop improvement for
sustainable agriculture / edited by M. Brett Callaway and
Charles A. Francis. p. cm. –
(Our sustainable future ; v. 4) Includes bibliographical
references and index.
ISBN 0-8032-1462-6 (cl) 1. Crop improvement.
2. Sustainable agriculture.
I. Callaway, M. Brett (Mitchell Brett), 1960–
II. Francis, Charles A.
III. Series. SB106.147C76 1993
631.5′23–dc20
93-18567
CIP

Contents

Contents

Contents

Contents

9. Contributions of Biotechnology to Crop Improvement, 157

Susan R. McCouch, Pam Ronald, and Molly M. Kyle

10. Genotype by Environment Interaction in Crop Improvement, 192

Ed Souza, Jim R. Myers, and Brian T. Scully

Contents

. . .

Tables and Figures

Figures

Tables

Preface

Crop improvement is fundamentally important to the continued productivity and economic viability of all agricultural cropping systems. Crop improvement for *sustainable* agriculture implies a change in emphasis given to specific breeding objectives rather than the development of a totally new set of crop improvement methodologies. In general, more emphasis is placed on increasing diversity at all levels within the system, from the deployment of individual genes to crops and cropping environments. Other objectives receiving greater emphasis are improvements in stress tolerance and resource use efficiency. In general, the future focus will be the adaptation of crop plants to systems and prevailing conditions rather than the expensive modification of the field environment to fit current crops.

We acknowledge that crop improvement is only one necessary component of sustainable agriculture. Larger social, cultural, economic, and political issues have led to the current challenges that face agriculture and rural people. These factors will continue to influence the choice of practices and the prevalence of particular agricultural systems. Biological realities have too often been ignored and ecological systems dominated by the temporarily expedient use of large inputs of chemicals and fossil fuels. The interest in developing sustainable agriculture results in large part from a realization that we can build on natural biological systems and take advantage of their potential for cycling resources and producing crops efficiently.

Crop Improvement for Sustainable Agriculture brings together scientists from several disciplines whose experiences and research interests relate directly to improving the sustainability of agricultural systems through crop improvement. It is hoped that these perspectives will provide direction and stimulus to others wishing to improve the profitability, long-term viability, environmental soundness, and social stability of agriculture.

This book results from the efforts of many people. Discussions with Dr. Tom Barker (Pioneer Hi-Bred, International) early in the book's development were particularly stimulating. Special thanks are due to Dorothy Callaway for proofing every chapter. Many colleagues contributed their time by reviewing one or more chapters. Their efforts are gratefully acknowledged. Final thanks are expressed to our families for their patience, support, and encouragement.

1

Crop Improvement for
Future Farming Systems

Charles A. Francis and M. Brett Callaway

Plant breeders are making decisions today that will influence the genetic contributions of crops to the farming systems that will prevail into the early 21st century. Because of the lead time involved in the genetic manipulation of plants, it is critical to envision the nature of those systems and the demands they are likely to place on future varieties and hybrids. Biotechnology or genetic engineering techniques will provide some new efficiencies in certain steps of the plant breeding process; but the wide testing, seed increase, and validation with farmers that must occur will still require time. Contributions to productivity through plant breeding result from years of investment and hard work.

Plant breeders will have to adjust selection criteria and program priorities to the new realities of changing farming systems and the broader decisions by society that impact agriculture. For example, genetic resistance to insecticides in some insect species, or government regulations on products, may preclude control with chemicals, thus forcing the use of more complex rotations or other types of biological management. Costs of fossil fuels may change the relative energy investment needed to manage weeds with herbicides versus cultivation. Perceptions of risk from some products used in agriculture may cause greater regulation and changes in farming practices. By anticipating broader changes in resources, climate, environmental regulations, and consumer perceptions and demand, we can start to sort out the many possible future scenarios and focus on those most likely to occur. Our predictions will not be perfect, but the alternative is to avoid any vision at all, thus missing out on opportunities that become available. Later in this chapter, an approach to predicting future systems will be presented, as well as an alternative and more proactive approach for researchers or development specialists to follow.

Although it is difficult to predict the precise nature of change, there are many indicators that can be integrated into a likely projection of the pressures that are going to impact food production. How we react to those pressures could be described as facing a crossroads and time of decision in agriculture (Coffman and Bates, this volume, Chapter 2). One path is to fine tune current systems, solving the most obvious problems and supporting the improvement of our current highly productive agricultural systems. The other is to consider carefully the challenges to these systems—high fossil fuel resource use, negative environmental impacts, inequitable social benefits—and design new systems that are biologically, economically, and socially sound. According to Coffman and Bates (this volume, Chapter 2), this new agriculture would "provide for a continued, balanced cycling and replacement of resources on a worldwide scale and do so in a socially responsible and equitable manner." In all likelihood we will do some of both, fine tuning current systems while looking down the road for major changes that will help alleviate the principal problems generated by today's agricultural systems.

The role of plant breeding is likely to be large in the design of future systems, but it is important to compare the likelihood of success through the genetic manipulation of germplasm to solutions that are provided by other changes in practices or crop combinations and sequences. Do we have the germplasm resources needed to provide adaptation to new systems or cropping conditions? Plant breeders will be looking for specific traits such as tolerance to pests, resistance to greater stress in some systems, and changes in crop quality for new uses. There are some new breeding technologies that will change the relative probabilities of success of finding or incorporating certain genetic traits in useful agronomic cultivars. A wider participation by farmers in the research and testing process provides the potential for developing and identifying specific adaptation to sites and systems. This chapter reviews some of our perceptions of the future, how these will be shaped by changes over the next decade, and how crop improvement will change. The recent book edited by Sleper et al. (1991), *Plant Breeding and Sustainable Agriculture: Considerations for Objectives and Methods,* should be consulted for additional ideas about future directions in crop improvement; the book includes a chapter on future contributions from plant breeders (Francis, 1991).

What Is Sustainable Agriculture?

Major quests in agriculture today involve seeking greater resource use efficiency, less negative impact on the environment, improved food safety and quality, and long-term profitability. Although not all people embrace the term, these goals are often found in the definitions of *sustainable agriculture.* The most prevalent components of these definitions are reviewed in order to provide a framework for the chapters that follow.

There is a wide range of opinions on how to define sustainable agriculture, with differences depending on philosophy, time and space reference, assumptions about the future, and even individual personality. Although it is not profitable to spend too much time on definitions, it is important to recognize that words and terms have meaning and that to communicate with others we each need to share that meaning and find common frames of reference. *Sustainable agriculture* could be considered a series of farming practices, a strategy for farming, a goal to be pursued, or a philosophy used to set goals for the long term. Definitions were reviewed by Francis and Youngberg (1990) in a recent book on sustainable agriculture (Francis et al., 1990); in the same volume Lockeretz (1990) described some of the major unresolved issues that cause misunderstanding in publications and discussions on this topic:

• Sustainable for how long?

• Sustainable under what assumptions about resources, environmental impact, or global population?

These are some of the complications that surround the term.

A short sample of definitions will help to focus this chapter and the rest of the book. Edwards (1988) stated that sustainable agriculture involves "integrated systems of agricultural production less dependent on high inputs of energy and synthetic chemicals and more management-intensive than conventional monocultural systems. These systems maintain, or only slightly decrease, productivity, maintain or increase net income for the farmer, are ecologically desirable, and protect the environment." In a more narrow context, Hoeft and Nafziger (1988) suggested that sustainable agriculture could best be defined or achieved by systems that were designed for maximum economic yield. In another symposium, Harwood (1990, p.4) described "an agriculture that can evolve indefinitely toward greater human utility, greater efficiency of resource use, and a balance with the environment that is favorable both to humans and to most other species." Wendell Berry suggested that the most desirable systems are those that degrade "neither land nor people" (from Coleman, 1987, p.x). No wonder there is difficulty in communication when people gather to discuss sustainable agriculture!

A review of a large number of such definitions reveals some elements that occur frequently. These are:

• Resource efficiency: systems that make the most efficient possible use of nonrenewable resources and whenever possible substitute renewable or internal resources for those from outside the farm.

• Profitability: systems that are economically profitable in both the short and the long term (if it's not profitable, it's not sustainable).

• Productivity: systems that build the soil and generally enhance the productivity of this basic resource instead of degrading it.

• Environmental soundness: systems that have a minimal negative impact both on the farm and beyond the farm borders, often measured in terms of water or air quality.

• Social viability: systems that are equitable, favor the success of owner/operator farms, encourage entry level farmers, and contribute to a viable rural economy, infrastructure, and community.

Few published definitions include all of these elements, but this list helps to bring focus to what is being discussed in the area of sustainable agriculture. From one perspective there is no disagreement: no one advocates a nonsustainable agriculture, although many would be quick to debate any specific definition. A summary definition was presented by Francis and Youngberg (1990, p.8) in their introductory chapter to a recent book:

> Sustainable agriculture is a philosophy based on human goals and on understanding the long-term impact of our activities on the environment and on other species. Use of this philosophy guides our application of prior experience and the latest scientific advances to create integrated, resource-conserving, equitable farming systems. These systems reduce environmental degradation, maintain agricultural productivity, promote economic viability in both the short and long term, and maintain stable rural communities and quality of life.

Through the remaining chapters different definitions or emphases will be used by different authors. Although this may cause some confusion, the editors feel that this divergence of opinion adds to the variety and richness of the discussion. We hope that such diversity will add strength to the overall sustainability of ideas and approaches in plant breeding, just as biological diversity in the field adds greater stability to yields and profits.

Will Future Cropping Systems Be More Sustainable?

As Harwood observed in 1979, "The basic agronomic, plant, and animal sciences on which agricultural development most heavily depends do not define the goals that development should pursue." Other forces, particularly economic, political, and social forces, define the goals and to a large extent the future shape of agriculture. The biological principles upon which the agronomic, plant, and animal sciences were built largely have been ignored. Large infusions of inputs have made this temporarily possible. Many of these inputs are dependent not only on the availability of fossil fuels but also on the availability of *inexpensive* fossil fuels. This fact alone dictates that more sustainable alternatives to present agricultural systems are needed. When one considers the growing social and political concern about the environmental damage resulting from the excessive or improper use of

5

many of these inputs, however, the argument for sustainable alternatives becomes almost overwhelming. Francis (1991) listed a number of potential changes in agricultural systems that have a high likelihood of taking place over the next 10 to 20 years. They are summarized below (see Francis, 1991, for details on potential plant breeding solutions):

- Reduced pesticide inputs and more regulations.
- Higher energy costs and more regulations on groundwater.
- Reduced tillage and more crop residues.
- Higher costs of irrigation.
- Greater use of specific adaptation.
- Higher use of crop rotations.
- Regulation of erosion levels.
- Increased use of multiple cropping systems.
- More diversity of crops and products for global markets.
- Increasing concerns about crop nutritional quality.
- Need for multiple purpose crops and plant types.
- Need for new crops, especially perennials.

Most of these potential changes would reduce our dependency on relatively inexpensive fossil fuels and increase sustainability of agricultural systems. As Coffman and Bates (this volume, Chapter 2) remind us, however, sustainable agriculture is not possible unless human population is stabilized and economic demand is more representative of social demand.

Role of Plant Breeders in Crop Improvement
for Sustainable Agriculture

Until quite recently (in terms of human history), crop improvement was accomplished by farmers and gatherers of wild plants (Coffman and Bates, this volume, Chapter 2). Today, plant breeders are responsible for genetic improvement in most crops. Their ability to improve crops depends on the availability of genetic variation. Genetic diversity is a central theme in sustainable agricultural systems and is of particular interest to those breeding for future systems. Goodman (this volume, Chapter 3) discusses the importance of this variability and how it may be best used by plant breeders.

Seed companies play a major role in developing cultivars and providing seed to the farmer. This role is not likely to change in the foreseeable future. Duvick (this volume, Chapter 4) assesses the role of seed companies in crop improvement for sustainable agriculture and concludes that many of the needs of sustainable agriculture will be met through a continuation of current breeding efforts. Highly specialized or localized needs, or traits that are particularly difficult to breed for, may not be addressed by commercial seed companies. He suggests that small companies may be more likely to develop cultivars for specialized or localized niche markets.

We feel that plant breeders at public institutions should assume a large responsibility for developing cultivars to meet important yet specialized or localized needs. In these situations, traits having a low probability of providing financial reward may nevertheless be quite important for society. Examples include improvements in nutritional quality, resource use efficiency, or adaptation to regions that are relatively unimportant to seed companies in terms of market potential yet whose economies depend on certain crops. Another very important role for public plant breeders is germplasm adaptation. Private companies usually cannot afford the large investment of time and capital required to develop exotic germplasm into potentially useful forms. An example of a highly successful germplasm adaptation program is the sorghum conversion program conducted by the USDA and several cooperating institutions. In this program, photoperiod-sensitive tropical sorghum is converted to photoperiod-insensitive forms that are able to mature under temperate conditions thus making them useful to temperate-zone breeders (House, 1985). Largely as a result of this program, commercial temperate sorghum hybrids have had a larger infusion of exotic germplasm in recent years than other crops such as maize. Many other crops would benefit from similar work.

Crop Improvement: Approaches and Applications

Crop improvement for future systems will continue to emphasize many of the breeding objectives of historical importance, such as resistance to insects and diseases, while placing new emphasis on areas such as tolerance to

7

biotic and abiotic stress and the development of new crops. Which of these and many other objectives a particular breeding program should emphasize is of considerable importance and too often neglected. Francis (this volume, Chapter 5) discusses the considerations required for the sound choice of objectives and the choice of parents, breeding and selection methods, and testing environments to best reach the objectives. Resistance to insects and diseases in future cropping systems is an objective whose importance few would deny. In many cases of relatively simple inheritance breeding for resistance or tolerance has been successful. Where breeders have put priority on complete resistance to some organisms, however, there has been sufficient pressure on the populations of these organisms to change and for some to survive even on the resistant hosts. Continual population changes in wheat rust or sorghum greenbug provide examples of these genetic modifications by pests to survive in a new crop environment. Certain organisms will be less prevalent in a changed system, for example from monoculture to rotation or intercropping, while others will be more of a problem. This is where relative priorities may change in a breeding program. In general, the diversity of rotations or complex multispecies systems will provide some system-based protection against crop pests and will allow a lower priority to be placed on selection for resistance. Similar to breeding for monoculture, selection for tolerance rather than complete resistance may put less pressure on the pest populations and allow a more lasting economic solution through breeding. Specific examples of these approaches are given by Hoffman et al. (this volume, Chapter 6).

Tolerance to abiotic stresses will be of greater concern in future cropping systems. Success in breeding for other traits that provide genetic advantage in sustainable systems will depend on the heritability of a specific trait and the priority placed on that characteristic in a breeding program. Crop selection for stress tolerance has been successful in some directions. Cold tolerance early in the season, an ability to germinate and begin development, has been found in a number of crops. The ability to flower and set seed under much lower than optimum temperatures has been found in grain sorghum and pearl millet. Tolerance to frost at the other end of the season is more ephemeral. Drought tolerance breeding is a complex endeavor because of the unpredictability of time of onset and severity of the stress. Nevertheless, some genetic progress has been made in finding sorghum cultivars with a higher degree of drought tolerance than current commercial hybrids. Blum (1988) and Christiansen and Lewis (1982) are excellent references for details.

Tolerance to biotic stress, other than insects and diseases, primarily involves tolerances to neighboring plants and has been discussed in the context of multiple cropping systems by Francis (1981, 1985), Gomez and Gomez (1983), Smith and Francis (1986), Wein and Smithson (1981), and Willey and Rao (1981). Crop tolerance to associated noncrop plants—that is, weeds—however, has received little attention. Callaway and Forcella (this volume, Chapter 7) provide evidence that crop cultivars differ in their tolerance to weeds and suggest methods of selection for improved tolerance to weeds.

New crops will be used in future cropping systems for a variety of purposes. They may be particularly well suited to harsh environments in which high levels of inputs are currently required to grow available crops. Perennial crops are of particular importance to future cropping systems because soil disturbance and erosion are minimized in their culture. Publications by the National Academy of Sciences (such as 1975; 1979; 1980; 1983a,b,c; and 1984) have described a number of these potential crops. Multipurpose tree species have been the focus of intense interest by the National Academy of Sciences and other organizations. Dr. Brewbaker has been actively involved with multipurpose tree species for a number of years and assesses the potential for tree improvement for agroforestry systems (this volume, Chapter 8).

Other potential new crops deserving special mention are perennial cereals and legumes that could contribute to a perennial commercial analog of the native prairie system. Envisioned by Jackson (1985) and described in detail by Soule and Piper (1992), a system composed of these perennial species would combine multiple species of plants with differing life cycles, nutrient and water needs, and contributions to food and feed supply. For example, a mixture might include a warm-season grass, a cool-season grass, and a perennial legume to sponsor nitrogen for the system. One or more of these crops could be harvested for seed, while the entire mixture could be grazed during one or more seasons during the year. One advantage would be soil protection by a permanent ground cover and low-cost maintenance, and another would be the stimulation of the biological structuring and stability found in native plant ecosystems. According to reports from The Land Institute (Soule and Piper, 1992), the primary questions that must be answered before such a system could become a reality include:

• Can a herbaceous perennial seed crop yield as well as an annual crop?

• Can a polyculture of perennial seed crops outyield the same crops grown in monoculture?

• Can a perennial polyculture provide much of its own fertility?

• Can a perennial mixture successfully manage weeds, harmful insects, and plant pathogens with little or no human intervention?

Soule and Piper present an up-to-date summary of the work done at The Land Institute as well as in other research organizations to provide partial answers to these questions.

Plant Breeding Goals and Genetic Resources

A broad goal of plant breeding programs is the development of cultivars that provide the maximum genetic yield potential under a prevailing set of climatic or cultural situations. There are secondary goals such as grain or forage quality and crops for multiple uses, but grain or tuber yield is the most frequent criterion for evaluation by both breeders and producers. As researchers begin to focus on broader objectives such as developing cultivars for sustainable systems, will these criteria change? We will be asking questions such as:

• What maize yields are possible when the crop is grown simultaneously with a leguminous cover that competes for moisture?

• What are the consistent potential soybean yields that can be achieved when the crop is relay planted into growing winter wheat in areas with an average rainfall of 740 mm (30 in) per year and high variation from year to year?

• Can we develop a grain sorghum hybrid with white seed for human food use that also has adequate stalk quality for grazing livestock during winter months?

Plant breeders have always worked on simultaneous selection for a number of traits. They start with the best agronomic types available and try to incorporate a small number of additional traits from other lines or an exotic source. Backcrossing to the well-adapted parent helps to maintain agronomic adaptation while allowing selection among progeny for the new traits. Focus on the smallest possible number of traits leads to the greatest genetic progress for those traits, whereas the selection for multiple traits delays or limits the progress that is possible in most cases for any single trait. Thus the specific goals of a breeding program are not to be taken lightly, as this step will predetermine to some extent the genetic progress that is possible.

In a rapidly changing agriculture it is difficult to set goals that will remain viable from the time of the first crossing of parents until the release of new cultivars. The more generations of selection that take place, the more genetically narrow the emerging new candidate cultivars become and the more difficult it becomes to change the course of selection. Although some fine tuning to new conditions will be possible, depending on residual genetic variation in materials in progress, it will generally be necessary to follow through to completion using the original criteria for selection. The other courageous option is to abandon the project altogether if it becomes obvious that the end product will not fit a market. This is a difficult choice and one not often taken.

Some examples will be useful to put these difficult decisions in perspective. With fairly frequent changes in greenbug biotype, sorghum breeders in the Great Plains have been faced with the decision to continue advancing lines to produce sorghum hybrids resistant to the most prevalent biotypes today, or to begin breeding for resistance to emerging biotypes. Changes in the perception of a potential market provide another example. At some point in the development of hybrid wheats in the United States, several major companies decided that this was not a viable and profitable option. Breeding activities soon became concentrated in a few companies that still considered this venture a promising commercial approach. Far beyond the control of a plant breeder in the United States, the rapid expansion of the production of a new crop in some other state or part of the world may make local production of the same crop a nonviable enterprise, regardless of success in an on-going breeding program. The decision may be to abandon the program. These are expensive decisions but highly practical changes that must continuously be considered in the design and management of a crop breeding program. A number of the chapters that follow consider plant breeding goals in pursuit of specific traits, many of which will help equip new cultivars for more stressful conditions and contribute to the sustainability of systems.

Do we have access to the germplasm resources needed for successful adaptation to future systems? Goodman (this volume, Chapter 3) has worked for many years with the races of maize and is familiar with the crop genetic banks that store this vital resource for the major food crops. He is convinced that a wealth of genetic variability exists and that this resource will be useful in the selection process for lower input systems. Goodman further concludes that plant breeders should build on the past half century of selection for yield potential in current cultivars and should use this highly productive genetic resource as the major input for new cultivars selected for future systems. The genetic potential in current cultivars could be called the "pre-adapted variety" component of a long-term crop improvement strategy (Duvick, this volume, Chapter 4).

In some crops or associated species there is a wealth of untapped variability, as described by Brewbaker (this volume, Chapter 8) for tree and shrub species. Many of these species are targeted for special and multiple uses in a system. The indigenous relatives of most species have not even been collected much less screened or selected for use in multiple plant type cropping or agroforestry systems. Tree and shrub species have passed through countless generations of selection for survival and may be among the most sustainable species we have available around which to design future systems. The genetics of many species is complex, since there is a high level of variability within the populations and many are polyploids. Of all the germplasm resources available for future systems, woody plants probably offer the most variability and potential for selection in any desired direction. As with other components of the system it is important to design the total mix with clear goals in mind, and sustainable natural ecosystems that include perennial species may hold some clues for this design.

New or Modified Methods for Plant Breeding

There is debate about the need for new plant breeding methods that will make selection for sustainable systems more feasible or successful. Coffman and Bates (this volume, Chapter 2) suggest that new techniques are required "to increase breeding efficiency and thus reduce the cost of seed or other propagules." They argue that new methods for selection and statistical analysis will make the breeding process more rapid and efficient and will provide approaches that help plant breeders to keep up with the changing nature of agriculture and its component systems. In contrast, Francis (this volume, Chapter 5) concludes that current plant breeding methods and time-proven designs will be adequate for crop improvement in the future. Both agree that priorities in crop improvement will change and that the modifications in cropping systems will necessitate careful setting of new breeding priorities for crop cultivars developed for future systems. There is also agreement on the need for multiple testing locations and for placing those in physical and agronomic situations where they will be most similar to the systems in which the cultivars will eventually be used. Recent experience in the commercial seed industry shows that multiple locations are far preferable to the replication of experiments in a single or small number of locations in evaluating the range of adaptation of a cultivar and its yield stability relative to other cultivars.

Biotechnology provides a number of tools that are based on basic advances in molecular biology. Some examples are discussed by McCouch et al. (this volume, Chapter 9). They describe the two major contributions of this science to be first, the increase in the efficiency of selection for some traits and second, the generation of new combinations of genetic variation. Since there is a large storehouse of genetic variation already available for most crops in the international centers and the global germplasm system, it is likely that the increased efficiency brought by some biotechnological

13

techniques will have the greatest potential impact on cultivar development for sustainable systems.

Other important tools are the statistical procedures that allow us to better quantify genotype by environment (GxE) and genotype by system interactions (Souza et al., this volume, Chapter 10). Calculation of GxE interactions helps our understanding of the importance of change in genotype as a component of systems, at least for the systems included in the available array of testing sites. Souza and colleagues distinguish among a number of different types of stability, the "reliable performance of a cultivar over a range of environments." How important stability will be depends on the crop, the other component crops in a system, and the relative values placed on component crops and risk associated with fluctuations in their yields. Several methods of analysis of GxE interaction are presented with examples in Chapter 10.

When crops are grown in more complex systems, and these are part of a strategy to achieve greater stability or sustainability of yields or income, there is need for more sophisticated statistical analysis. Federer (this volume, Chapter 11) describes the design of experiments to study intercropping species combinations, from the most simple of two-crop mixes to those involving more species in the same field. Although his methods and the majority of publications to date have dealt with crop yields and less frequently with net income, there is an important and broader future dimension to sustainability. As Federer points out, there may be greater interest in other measures of success, including land use efficiency, nutritional productivity, reduction of soil or other resource loss, or tolerance to pests without application of pesticides. There may be interest in net return at minimal risk to other limiting factors such as energy, labor, water, or days of growing season. All of these research priorities will require careful choice of field methods and designs, approaches to statistical analysis and evaluation of data, and interpretation of results in terms of the factors deemed most important by producers. Some methods may be new, and many are likely to be those that have proved valuable to plant breeders over the past century. If the methods do not change drastically, what of the way in which crop breeders perceive the future? In fact, plant breeding is an activity and a discipline that by definition must be working for the future.

International Dimensions of Crop Improvement

Future systems will be shaped by biological, economic, environmental, and social changes that occur globally. Many of these changes are beyond the reach and even the comprehension of most scientists, yet they influence the course of our work. Farmers' success is also greatly impacted by weather events and changes in production practices in other countries and by political negotiations and international markets. It is difficult to imagine how an individual scientist or farmer can adjust a cropping system or change a crop cultivar to make any difference when confronted with these global changes or influences. We have generally viewed the future as a series of circumstances and constraints to which we must adapt our cultivars. This is not the only option.

There is another way to view the future, and it is a more positive and empowering view. Rather than letting the future happen to us and to our predominant cropping systems, we can analyze events and directions and essentially plan a more desirable future. This can be done for an individual farm, a watershed, a community, or a country. The essence of this process is to envision a most desirable place and set of circumstances under which we would want to live, in concert with resources and a habitable environment, and then gain consensus with others on how to design that future. The next important step is to start making decisions today that will cause that more desirable future to happen. In cropping, the essence would be a sustainable set of systems that make maximum possible use of renewable resources; cause minimal or no deterioration of soil, water, or the rest of the environment; and reach these goals in a socially equitable manner. With agreement on goals, dedication of enough resources, and use of the genetic potential of current and new crop species, we can envision and reach such a future. As outlined in the chapters that follow, plant breeding has the potential to continue to make a major contribution to the sustainable production of crops in the future.

References

Blum, A. 1988. Plant breeding for stress environments. CRC Press, Boca Raton, FL.

Christiansen, M.N., and C.F. Lewis (ed.). 1982. Breeding plants for less favorable environments. Wiley, New York, NY.

Coleman, B. 1987. Preface and acknowledgments, pp.ix–xi. *In* W. Jackson, W. Berry, and B. Coleman (ed.) Meeting the expectations of the land. North Point Press, San Francisco, CA.

Edwards, C.A. 1988. The concept of components of sustainable agriculture. p.39–41. *In* C.A. Francis and J.W. King (ed.) Sustainable agriculture in the Midwest. Proc. North Central Regional Conf., Agr. Res. Div. and Coop. Ext. Service, Univ. Nebraska, Lincoln, NE.

Francis, C.A. 1981. Development of plant genotypes for multiple cropping systems. p.179–231. *In* K.J. Frey (ed.) Plant breeding symposium II, Iowa State University Press, Ames, IA.

Francis, C.A. 1985. Variety development for multiple cropping systems. CRC Crit. Rev. Plant Sci. 3:133–168.

Francis, C.A. 1991. Contributions of plant breeding to future cropping systems. p.83–93. *In* D.A. Sleper, T.C. Barker, and P. Bramel-Cox (ed.) Plant breeding and sustainable agriculture: Considerations for objectives and methods. Crop Sci. Soc. Am. Spec. Pub. 18. Amer. Soc. Agron., Madison, WI.

Francis, C.A., C.B. Flora, and L.D. King (ed.). 1990. Sustainable agriculture in temperate zones. Wiley, New York, NY.

Francis, C.A., and G. Youngberg. 1990. Sustainable agriculture—an overview. p.1–23. *In* C.A. Francis, C.B. Flora, and L.D. King (ed.) Sustainable agriculture in temperate zones. Wiley, New York, NY. 1990.

Gomez, A.A., and K.A. Gomez. 1983. Multiple cropping in the humid tropics of Asia. Pub. 176e. IDRC, Ottawa, Canada.

Harwood, R.R. 1990. History of sustainable agriculture. p.1–19. *In* C.A. Edwards, R. Lal, P. Madden, R.H. Miller, and G. House (ed.) Sustainable agricultural systems. Soil Water Conserv. Soc., Ankeny, IA.

Harwood, R.R. 1979. Small farm development. Westview Press, Boulder, CO.

Hoeft, R.G., and E.D. Nafziger. 1988. Sustainable agriculture. p.7–11. *In* R.G. Hoeft (ed.) Proc. Illinois Fertilizer Conf., Dept. Agron., Univ. Illinois, Urbana, IL.

House, L.R. 1985. A guide to sorghum breeding. 2nd ed. International Crops Research Institute for the Semi-Arid Tropics, Andra Pradesh, India.

Jackson, W. 1985. New roots for agriculture. New ed. University of Nebraska Press, Lincoln, NE.

Lockeretz, W. 1990. Major issues confronting sustainable agriculture. p.423–438. *In* C.A. Francis, C.B. Flora, and L.D. King (ed.) Sustainable agriculture in temperate zones. Wiley, New York, NY.

National Academy of Sciences. 1975. Underexploited tropical plants with promising economic value. National Academy Press, Washington, DC.

National Academy of Sciences. 1979. Tropical legumes: Resources for the future. National Academy Press, Washington, DC.

National Academy of Sciences. 1980. Firewood crops: Shrub and tree species for energy production. National Academy Press, Washington, DC.

National Academy of Sciences. 1983a. Firewood crops: Shrub and tree species for energy production. vol. 2. National Academy Press, Washington, DC.

National Academy of Sciences. 1983b. Mangium and other fast-growing acacias for the humid tropics. National Academy Press, Washington, DC.

National Academy of Sciences. 1983c. Casuarinas: Nitrogen-fixing trees for adverse sites. National Academy Press, Washington, DC.

National Academy of Sciences. 1984. Leucaena: Promising forage and tree crop for the tropics. National Academy Press, Washington, DC.

Sleper, D.A., T.C. Barker, and P.J. Bramel-Cox (ed.). 1991. Plant breeding and sustainable agriculture: Considerations for objectives and methods. Crop Sci. Soc. Am. Spec. Pub. 18. Amer. Soc. Agron., Madison, WI.

Smith, M.E., and C.A. Francis. 1986. Breeding for multiple cropping systems. p.129–249. *In* C.A. Francis (ed.) Multiple cropping systems. Macmillan, New York, NY.

Soule, J.D., and J.K. Piper. 1992. Farming in nature's image. Island Press, Covelo, CA.

Wein, H.C., and J.B. Smithson. 1981. The evaluation of genotypes for intercropping. p.105–116. *In* ICRISAT. Proc. Int. Workshop on Intercropping, Hyderabad, India. 10–13 Jan. 1979. International Crops Research Institute for the Semi-Arid Tropics (ICRISAT), Patancheru, India.

Willey, R.W., and M.R. Rao. 1981. Genotype studies at ICRISAT. p.117–127. *In* ICRISAT. Proc. Int. Workshop on Intercropping, Hyderabad, India. 10–13 Jan. 1979. International Crops Research Institute for the Semi-Arid Tropics (ICRISAT), Patancheru, India.

2

History of Crop Improvement
in Sustainable Agriculture

W. Ronnie Coffman and David M. Bates

Harvesting sunlight to sustain life on earth is the vital role of plants, yet humans seldom think of plants as our only means of capturing energy for food or as the primary source of fossil fuels. Fossil fuel reserves, however, are finite and in the totality of the earth's energy budget are minor contributors. They represent the equivalent of only about one week of solar radiation. Just as selection and breeding have met past challenges in increasing plant productivity, new challenges posed by concerns for long-term sustainability and more effective use of the sun's energy through plants await human solutions. Increasing the utility and efficiency of plants to capture energy and convert it into food, fiber, and fuel is key to sustaining agriculture and human existence in the years ahead.

Crop improvement, or plant breeding, as broadly defined, is the genetic modification of plants. It includes a range of activities from molecular genetics to applied selection and testing by farmers. The improvement of plants through breeding has been responsible for most of the gains in production agriculture. In the modern era, economists (Arndt and Ruttan, 1977; Evenson, 1991; Ruttan, 1982, 1987) have shown that more than 50% of the gains are attributable directly to genetic improvement. Although data are not available to partition earlier gains, clearly the selection and maintenance of productive cultivars have been the most important activity in agriculture.

Some 10,000 years after the birth of agriculture, we stand at a crossroads. Despite tremendous contributions of plant breeding to the betterment of humankind, valid questions concerning the sustainability of current modern agricultural approaches have been raised. We know that deserts now exist where civilizations once flourished. Operating on a global scale, as we do now, we are vulnerable to global failure. We can continue to follow past pathways, or we can boldly seek new ones which will lead to sustainable agricultural systems. Such systems will provide for a continued, balanced cycling and replacement of resources on a worldwide scale and do so in a socially responsible and equitable manner.

In this chapter we examine some historical aspects of plant breeding as a foundation for expressing more contemporary concerns, paying particular attention to the recent impact and apparent limitations of the Green Revolution and offering suggestions for an approach to agricultural improvements that may be sustainable.

Plant Breeding and the Beginning of Agriculture

Agricultural origins in the Near East extend back in time at least 10,000 years. In other parts of the world their appearance was apparently somewhat later. In each instance the recovery of preserved plant materials showing characteristics of human selection provides the crucial evidence for agriculture. In the Near East the archaeological presence of nonshattering rachises of barley and wheat, both einkorn and emmer, signal agriculture. Other evidence, including nondehiscent legumes and seeds larger than those of wild types in peas, lentils, chickpeas, and bitter vetch supports the conclusion. Elsewhere in the world the archaeological recovery of plant remains showing similar fruiting changes, for example, those in the foxtail and broom millets in temperate China, rice in Southeast Asia, and in a constellation of crops including beans, squashes, and eventually maize in South and Meso-America, are among the crucial discoveries (see Harris and Hillman, 1989, for relevant references).

In each of the foregoing situations, the preserved plant remains are indicative of agricultural systems based on the sowing and harvest of seeds. Evidence for agricultural systems based on vegetative propagation, that is,

20

vegecultures, does not appear until later. This may be attributed to the absence of these systems in earlier times or to the poor potential for preservation of herbaceous plant parts. Lacking evidence, we do not know if vegeculture systems predated, were contemporaneous with, or were later than seed-based agricultural systems.

Although we know a good deal about where and under what environmental conditions seed-based agricultural systems first appeared, a definitive explanation of why agriculture began remains to be presented and perhaps never will be. The question, however, does not suffer from a lack of answers. Some, especially those that see in agricultural origins a cause and effect relationship, are deterministic in nature. They ascribe its beginnings to environmental, cultural, or populational causes. Other explanations, either singly or coupled with deterministic or other views, view agriculture as the outgrowth of environmental manipulation or coevolutionary events. Still others attribute agricultural origins to more esoteric events, for instance the evolution of cities or sacrificial or fertility rites. Summaries of these views and permutations of them may be found in Cohen (1977), Harris and Hillman (1989), Heiser (1990), Reed (1977), and Rindos (1984), among others. Those who seek a universal explanation for the human transition from hunting and gathering to agriculture, particularly when seeking to explain the essentially simultaneous appearance of agriculture in the Old and New worlds, tend to be drawn to deterministic camps. There is no reason *a priori* for a single explanation of agricultural origins. Most hypotheses hold certain truths, which have relevance under given circumstances, especially when placed in a holistic framework.

Whatever the underlying rationale for the origin of agriculture, its emergence was a revolutionary event in human history. In part, it replaced hunting and gathering systems, and in part it represented a fundamental change in the relationships of humans with nature. Hunter-gatherers exploit the resources of their surroundings in a seasonal cycle. Agriculturists modify and manipulate the land and plants of their environment to suit specific purposes. The cultivation of crops and the husbandry of animals permitted humans to harvest energy stored in biomass more productively and efficiently than is possible in most hunting-gathering contexts. In this sense agriculture has a selective advantage over hunting-gathering, one outcome of which has been the evolution of complex societies (Bates, 1985).

Emerging from a gathering and hunting background, early agriculturists apparently focused on plants that were already being gathered and were amenable to human selection. Thus, agriculture and plant breeding seemingly share a common origin, although the cultivation of plants was certainly a possibility before domestication took place. Selection, however, is the principal method of crop improvement. Without selection, coupled with an agricultural setting, the creation of domesticates would not have occurred. Other factors also played a significant role in early selection, and at least two seem to have been essential in seed-based systems: nonshattering rachises and annual habit. The retention of fruits in the infructescence permitted the efficient harvest of the mature grains and a means of perpetuating plants with desirable traits. The annual habit permitted the recurring cultivation of large populations from which the selection of desirable segregates was facilitated. In regions of the world where these conditions were not exploited, seed-based agriculture appears not to have developed until later in agricultural history.

Agricultural Transitions

Since our agricultural beginnings, successful plant breeders have responded to meet human needs. First were farmers who domesticated and selected the basic array of our food and fiber crops and later were professional plant breeders. The outgrowth of the early millennia of agriculture and selection by what we now refer to as traditional farmers was the creation of an astonishingly rich array of domesticated plants, encompassing both species diversity and a vast inventory of land races and genetic types.

In early agricultural periods plant diversity was high, for it reflected the peculiarities and diversity of environments and cultures and the isolation of peoples in different regions of the world. Plant species and their variants were selected not only for their productivity but also for their ability to yield in a predictable and secure fashion. As nations and economies of the world became more internationalized, especially over the past several decades, selective forces broadened and intensified simultaneously. Productivity rather than security became the driving force for selection. Coupled with subsidies offered through a wide range of direct and indirect government sponsored programs the result has been to concentrate world-wide subsistence on a few staple crops (Fig. 2.1) that now constitute humankind's primary, though not exclusive, subsistence base (Bates, 1985, 1987). Prescott-Allen and Prescott-Allen (1990) point out, as did Bates (1985, 1987) in the context of secondary use pools, that a much wider range of food plants actually constitutes the subsistence base. These also are coming under increasing selection pressures that are likely to reduce their numbers on a world-wide scale. Accompanying the narrowing of agricultural biodiversity has been a simplification of agricultural environments and expansion of homogenized systems throughout the world. The latter has been accomplished through high capital expenditures and the use of fertilizers, herbicides, pesticides, and other means.

Among food crops, over 50% of the food supply comes from seven cereal grains, over 40 percent from rice and wheat. This extreme concentration of the subsistence base is a result, in part, of the merits of successful crops and, in part, from an accounting system that does not recognize the full costs of producing these crops. The latter point is illustrated by plotting energy inputs against outputs as agriculture has evolved (Fig. 2.2). When we in the United States invest 12,000 calories of energy per day per capita to provide each American with the 3,300 calories that he or she on average consumes (Dahlberg, 1979), it is clear that present agricultural approaches and narrow crop focus are not sustainable indefinitely. The problem, of course, is more difficult for developing regions of the world, which lack the capital to subsidize lavish agricultural practices.

The Green Revolution

Is it just to criticize the Green Revolution, with its recognized accomplishments, for failure to correct all the social-economic ills of the world that have accumulated from the days of Adam and Eve up to the present?—Borlaug, 1972

For plant breeders the so-called Green Revolution was but a logical next step in the continuing development of wheat and rice cultivars designed to meet the demands of producers and consumers. Advances in water control

23

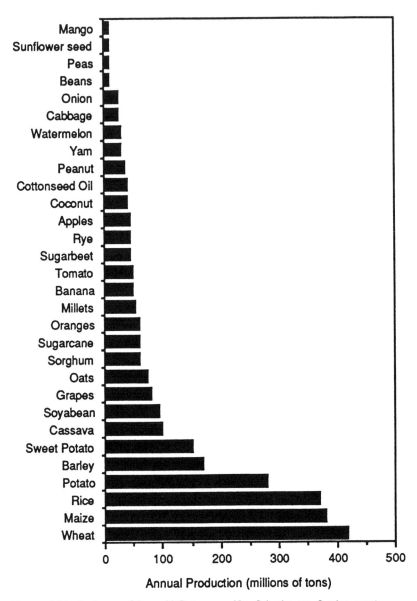

Fig. 2.1. Major food crops of the world (Data source: New Scientist, UK, October, 1990).

24

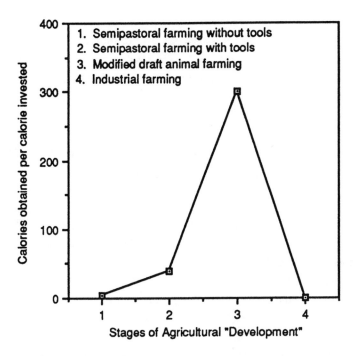

Fig. 2.2. Calories obtained per calorie invested at various stages of agricultural development (Data source: Cox and Atkins, 1979).

and fertilizer technology had created some relatively homogeneous growing areas for rice and wheat throughout the tropical world. Investments in plant breeding programs by the Rockefeller and Ford foundations in Mexico (wheat) and the Philippines (rice) produced short-stature, daylength insensitive cultivars widely adapted within irrigated areas with good water control. They were responsive to (not to say dependent on) the application of chemical fertilizers. Although not without its critics (Shiva, 1991), the impact of the Green Revolution was substantial and by most analyses largely positive.

The Green Revolution has been our best (not to say perfect) effort to meet the social demand for food. The real price of rice today is less than one-half of what it was 30 years ago when the International Rice Research Institute

Table 2.1. Production and area of wheat in India.

Year	Production (million tons)	Area (million ha)
1964	12	14
1965	55	23

Source: M.S. Swaminathan, personal communication.

launched its research program. This lower price has increased the availability of rice to lower income groups, benefiting millions of urban consumers. Farmers who cultivated the new varieties benefited considerably, particularly in the early stages of the revolution when production increased dramatically and the price was still high. Laborers employed by those farmers also benefited. The big losers were farmers who could not use the new varieties for one reason or another, usually poor water control. For them, production has remained stagnant while the price has dropped dramatically. In general, farmers in less-favored production areas have been disadvantaged because of the impact on other commodities of the low prices for rice and wheat. This has increased the pressure on the more marginal and fragile ecologies as people struggle to survive.

Often forgotten is the fact that the Green Revolution mitigated the large-scale diversion of forests and other marginal and fragile land for the cultivation of food. In the case of India (Table 2.1) the production of wheat has expanded more than fourfold since 1964, while in the same period the area given to wheat expanded less than twofold. Producing the same amount of food by expanding area alone, without any increases in yield, would have resulted in an environmental disaster.

Perhaps the greatest biological threat to sustainability brought about by the Green Revolution is the relative uniformity of the germplasm of major crops now deployed throughout the world. Hargrove et al. (1980, 1988) document this phenomenon and discuss its implications.

Breeding for Sustainability

Agronomic, social, nutritional, and economic justifications exist for en-hancing the world's agricultural potential through the maintenance of a biologically diverse crop base rather than one narrowed to a few primary staples. By no means can we forego the products of industrial agriculture, but we must modify it in ways that make it environmentally more benign and resist the temptation to extend it into fragile environments.

Breeding for sustainability is a process of fitting appropriate crops and cultivars to an appropriate environment instead of altering the environ-ment through inputs such as fertilizer, water, and pesticides. Many of the breeding objectives pertaining to sustainable agriculture come under the heading of stress tolerance. These include biological stress (diseases, in-sects, weeds, and other crop species), physical stress (drought, heat, and cold), and chemical stress (adverse soils). In sustainable systems where intercropping may be important, stress may be caused by other species of productive plants competing for limited light, water, and nutrients.

Many of the traits important to sustainable agriculture are complex and are difficult and expensive to select and evaluate. New techniques are needed to increase breeding efficiency and thus reduce the cost of seed or other propagules. Development of appropriate selection strategies, statisti-cal designs for testing, and breeding methods constitutes a portion of the basic plant breeding research needed to develop germplasm for sustainable agriculture. Application of these new techniques to appropriately chosen breeding objectives should improve the rate of development and quality of products from breeding programs. Bates (D. M. Bates, unpublished manu-script) makes the point that agriculture is neither monolithic nor static. Continuing changes in the state of humanity and the world dictate the con-tinuing evolution of agriculture along several interrelated tracks, each even-tually defined by the resources available to support it. Sustainable, low-input systems that combine traditional knowledge of plants and practices with that of modern biology will emerge and contribute in positive ways to the lives of the people who practice them. Bates emphasizes that the use of a wide array of crops will contribute to the overall agricultural enterprise by

• Increasing the extent of land that can be brought into production.

• Permitting the intensification of production in a given area.

• Extending the season of harvest, especially into periods unfavorable for the growth of nonadapted exotics (such as dry or wet periods), and providing a margin of safety against catastrophic loss of exotic crops.

• Creating ecologically diverse, stable, harvestable habitats in rural areas and, with appropriate technology, in urban areas as well.

• Providing a means of conserving germplasm and assuring continuing access to it.

• Lowering the costs of conventional agricultural inputs, such as pesticides, herbicides, and water, by using plants better adapted to local conditions.

• Creating a wider array of desirable foodstuffs and enhancing nutrition not only of humans but also, through fodder or feed, that of animals.

• Contributing opportunities to gain cash income and to foster improved standards of living for rural peoples.

Scientists involved in plant improvement have only begun to tap the potential of the vast plant genetic resources of the earth and the environments in which they could be used. Full use of this potential cannot be achieved unless institutions in the United States and elsewhere redefine their priorities and programs to attack the real problems that face humankind.

Institutional Roles in Crop Improvement

Coffman and Smith (1991) review the roles of public, industrial, and international research center breeding programs in developing germplasm for sustainable agriculture. The state of research at public institutions in the United States is summarized aptly in a new report from the National Research Council entitled "Alternative Agriculture" (Anonymous, 1989). The results and design of basic, discipline-oriented research programs are often not sufficiently integrated into practical interdisciplinary efforts to understand agricultural systems and solve some major agricultural problems.

Most public institutions, and many applied plant breeders in these institutions, have lost control of the research agenda. Through innovative grantsmanship and creative financial management some public institutions have been able to keep their applied programs intact while building capability in the emerging technologies. Public plant breeders, working with colleagues in associated disciplines, have been responsible for producing much of the technology that now exists for minimizing the adverse effects of agricultural production on the environment. These few public institutions are now in a position to produce the technologies increasingly demanded by farmers and others adversely affected by the steady intensification of agricultural production. In many public institutions, however, applied plant breeding capability and programs will need to be rehabilitated.

Industry is often blamed unfairly for many of the problems that we face in agriculture. We recognize that private industries must make a profit to survive, but we are sometimes dismayed when faced with the reality of decisions based on profit motives. It is clear that financial incentives exist for the development of hybrid cultivars that produce higher yields. But there is little incentive and a great deal of risk associated with producing cultivars designed to reduce input use and enhance alternative farming practices. It is not realistic to expect that industry will take those kinds of risks. Johnson (1984) understated the case when he noted, "Not all publicly desirable technologies will turn out to be privately profitable." He went on to assert that if privately unprofitable but socially desirable technologies are to be developed and used, it will be necessary for the public to support their development and distribute them at least to where it is privately advantageous for agribusiness to take them over and for farmers to use them. Private industry can be expected to invest in breeding for low-input sustainable agricultural systems only to the extent that it will make their products more competitive and help them to realize a profit. It seems clear that they will concentrate on producing hybrid cultivars with resistance or tolerance to biological, physical, and chemical stresses (Duvick, this volume, Chapter 4).

Buying Time with Food Stamps

An international food stamp program (Peterson, 1991) could augment the purchasing power of the world's poorest nations while stimulating the demand for agricultural commodities in the high-income, food surplus nations. The program could mitigate hunger and malnutrition in the Third World, rejuvenate rural economies in both developed and less developed countries, and stimulate world economic growth. Peterson believes that this could be accomplished for about the same amount of money now being spent by the world's high-income nations on farm income support and foreign food aid programs. Properly administered, it would place purchasing power in the hands of poor people and promote sustainable development at the local level in the developing world.

Lessons from History

Agriculture cannot be sustained unless population is stabilized. Policies must be adjusted on a global basis to bring economic demand in line with social demand so that plant breeders can contribute more effectively to the real needs of the world. Half the world's people are hungry, and they are not going to improve themselves substantially while they are in that condition. In fact, they are more likely to destroy the environment out of sheer desperation. We need an effectively managed global food stamp program (Peterson, 1991), the removal of all subsidies for water and transportation in developed countries, and the strict regulation of chemical inputs that may adversely affect the environment. Public institutions in the United States must provide the foundation for future breeding activities in support of low-input sustainable agriculture. Underutilized crops for underutilized environments, with due regard for the preservation of native biodiversity, should be the general research focus. Only well-funded research institutions in the public sector can be expected to act in the public interest. Industry deserves the support of these institutions to take the big risks in applied research that will provide future varietal technology. It is now clear that agricultural activities in the Third World will affect the future of the entire planet and that institutions in these regions will look toward public institutions in the United States and other developed countries as collaborators to conduct the "upstream" research needed to develop suitable technologies for Third World farmers.

As Harlan warned in 1975, "Research on unconventional foods will continue and many of the problems in production and utilization will be solved, but all of this striving may turn out to be fruitless. As long as the population of the world continues to rise indefinitely, no rational solution is possible. There is no procedure, process, or technique that will prevent eventual mass starvation unless the world population is stabilized."

References

Anonymous. 1989. Alternative Agriculture. National Academy Press, Washington, DC.

Arndt, T.M., and V.W. Ruttan. 1977. Valuing the productivity of agricultural research: Problems and issues. p.3–25. *In* T.M. Arndt, D.G. Dalrymple, and V.W. Ruttan (ed.) Resource allocation and productivity in national and international agricultural research. Univ. Minnesota Press, Minneapolis, MN.

Bates, D.M. 1985. Plant utilization: Patterns and prospects. Econ. Bot. 39:241–265.

Bates, D.M. 1987. The potential for new crops. Angew. Bot. 62:31–40.

Borlaug, N.E. 1972. The Green Revolution, peace, and humanity. Speech delivered upon receipt of the 1970 Nobel Peace Prize. CIMMYT Reprint and Translation Series No. 3. Centro Internacional de Mejoramiento de Maiz y Trigo, El Batan, Mexico.

Coffman, W.R., and M.E. Smith. 1991. The roles of public, industry, and international research center breeding program in developing germplasm for sustainable agriculture. p.1–9. *In* D.A. Sleper, T.C. Barker, and P.J. Bramel-Cox (ed.) Plant breeding and sustainable agriculture: Considerations for objectives and methods. Crop Sci. Soc. Am. Special Pub. 18. Amer. Soc. Agron., Madison, WI.

Cohen, M.N. 1977. The food crisis in prehistory. Yale Univ. Press, New Haven, CT.

Cox, G.W., and M.D. Atkins. 1979. Agricultural ecology: An analysis of world food production systems. W. H. Freeman, San Francisco, CA.

Dahlberg, K.A. 1979. Beyond the Green Revolution: The ecology and politics of global agricultural development. Plenum Press, New York, NY.

Evenson, R.E. 1991. Notes on the measurement of the economic consequences of agricultural research investments. Paper presented at the Cornell International Institute for Food, Agriculture, and Development Workshop on Assessment of International Agricultural Research Impacts for Sustainable Development, Ithaca, NY. 16–19 June.

Hargrove, T.R., W.R. Coffman, and V.L. Cabanilla. 1980. Ancestry of improved cultivars of Asian rice. Crop Sci. 20:721–727.

Hargrove, T.R., V.L. Cabanilla, and W.R. Coffman. 1988. Twenty years of rice breeding. The role of semidwarf varieties in rice breeding for Asian farmers and the effects on cytoplasmic diversity. Bioscience 38(10):675–681.

Harlan, J.R. 1975. Crops and man. Crop Sci. Soc. Am., Madison, WI.

Harris, D.R., and G.C. Hillman. 1989. Foraging and farming: The evolution of plant exploitation. Unwin Hyman, London, England.

Heiser, C.B., Jr. 1990. Seed to civilization. 3d ed. Harvard Univ. Press, Cambridge, MA.

Johnson, G.L. 1984. Academia needs a new covenant for serving agriculture. Michigan State Univ. Press, East Lansing, MI.

Peterson, W. 1991. World hunger: A solution with food stamps. Choices 6(2):24–26.

Prescott-Allen, R., and C. Prescott-Allen. 1990. How many plants feed the world? Conserv. Biol. 4:365–374.

Reed, C.A. (ed.). 1977. Origins of agriculture. Mouton, The Hague.

Rindos, D. 1984. The origins of agriculture: An evolutionary perspective. Academic Press, New York, NY.

Ruttan, V.W. 1982. Agricultural research policy. Univ. Minnesota Press, Minneapolis, MN.

Ruttan, V.W. 1987. Toward a global agricultural research system. pp.65–97. In V.W. Ruttan and C. Pray (ed.) Policy for agricultural research. Westview Press, Boulder, CO.

Shiva, V. 1991. The Green Revolution in the Punjab. The Ecologist 21:57–60.

3

Choosing Germplasm for
Breeding Program Success

Major M. Goodman

Abundant evidence suggests that plant breeders can achieve gains from selection ranging from 1 to 4% per year with cross-pollinated crops such as maize (Hallauer and Miranda, 1981) and only somewhat less than that with self-pollinated crops such as wheat or soybeans (Fehr, 1987). Carefully tailoring breeding methods to the crop can optimize these rates, but a wide range of breeding methods works almost equally well (Moll, 1978; Bauman, 1981). Other factors are often more important to the success of a breeding program than the specific selection methodology followed; these include choice of germplasm, quality and extent of yield- and disease-testing programs, and skill of the breeder. It is widely perceived that the quality and scope of yield-testing programs largely determine the relative success of the major seed companies today. All have breeders with distinctive points of view using diverse arrays of materials, often following selection strategies unique to each location.

Nonetheless, history clearly suggests that a few breeders with unusual insight can have a major impact, even when testing and selection procedures may not be optimum. The records of breeders such as Raymond Baker (Pioneer Hi-Bred), Lester Pfister (Pfister Associated Growers), George Sprague (USDA and University of Illinois), and G.H. Stringfield (Ohio State University and Dekalb Agricultural Research) in maize; Norman Borlaug (International Maize and Wheat Improvement Center—CIMMYT) in wheat; and the horticulturist Luther Burbank (nurseryman) are ample evidence that an individual breeder can have a major impact on the success of a program.

There is some evidence that seed companies that allow individual breeders flexibility in breeding programs are, in the long run, more successful than those that dictate standardized breeding methods applied to specific populations. This evidence is largely based upon the past success of companies such as Pioneer Hi-Bred International that traditionally attempted to hire good people and let them run their own breeding programs. The success of flexible, multifaceted programs is probably preordained in the rapidly changing agricultural environment our farmers have faced in the past 75 years (McMillen, 1991). It is likely that change will accelerate for the next 75 years. The results of molecular biology will probably become commonplace, and petroleum-based fuels and agricultural chemicals are apt to become relatively more costly. These and other factors, many now unforeseeable, almost guarantee accelerated change.

Germplasm and Sustainable Agriculture

The selection of germplasm is critical to a breeder's success, yet it receives little attention in graduate or undergraduate education. Most often it is learned by apprenticeship on the job, often with costly consequences, since years may be required to learn that the materials initially chosen are inadequate for today's demands. In addition, the development and marketing of commercial cultivars usually involve a ten- to fifteen-year cycle, and demands can easily change during the ensuing decade(s). An inconsequential disease or insect can become rampant, or farming practices can require totally different harvest maturities, planting times, or fieldbed environments.

Today the breeder may be faced with developing cultivars for sustainable agriculture that must satisfy diversified needs ranging from Amish farmers using high inputs and much hand labor, maintaining extraordinarily weed-free fields, to dirt farmers who raise no livestock and whose only input other than seeds may be an occasional green manure crop such as alfalfa. In addition the breeder must also satisfy the more numerous moderate-to-high-input farmers. Today more consumers are demanding pesticide-free agricul-

34

tural products, but many currently grown cultivars are inappropriate for production without pesticides. Thus consumer choice is often between buggy, pesticide-free fruits and vegetables or attractive ones grown using pesticides. The sustainable agriculture movement has created a demand for which plant breeders will soon scramble to supply cultivars.

There is little doubt that today's cultivars have not been selected for low-input farming and are unlikely to be optimal for such use. These cultivars result, however, from thousands of years of selection by primitive agriculturists whose work probably accounts for most of our current agricultural productivity. The peasants who domesticated and developed our crop plants had to maintain sustainable agricultural productivity or return to hunting and gathering. In addition to today's cultivars, breeders have available to them abundant supplies of other germplasm that may be better adapted to lower inputs than are elite cultivars. It is likely that some of this germplasm, often elite or obsolete cultivars from elsewhere in the world, will make a substantial contribution to breeding programs in the future.

Alternative Germplasm: Not a Panacea

Before discussing germplasm choices and their likely contributions, let us first explore some pitfalls of the "germplasm solution" to minimum-input farming, lest they result in minimum output as well. First, although today's elite cultivars were not selected under minimum-input, minimum-tillage conditions, they are often much higher yielding, more stress tolerant and better adapted to modern tillage practices than the cultivars developed before the 1950s (Duvick, 1977; Russell, 1974; Castleberry et al., 1984; Meghji et al., 1984). Although different choices of breeding materials in the 1930s and 1940s might have resulted in today's cultivars being better adapted to alternative agricultural practices, turning back the clock and abandoning 50 years of intensive selection is not a viable alternative in most cases. The argument that the use of a different set of selection procedures during the past 50 years would have dramatically changed the status of today's cultivars is not a new one (Berlan and Lewontin, 1986). Although it has some rationale, it is unlikely to change current breeding methods and

35

materials. Given a 50-year head start, even a mediocre breeding program based upon good, but not optimal, procedures and germplasm will likely produce better, more widely adapted cultivars than a new program using optimal procedures but with germplasm lacking the 50 years of selection. Rankings of cultivars across a wide range of environments are often quite similar. For example, estimates of the correlation between maize testcross yields in North Carolina with those in Sao Paulo, Brazil, were only slightly smaller than correlations across different testers in North Carolina (0.31 vs. 0.33, respectively; Moreno-M., 1989). Thus selection conducted under one set of conditions, such as high-input farming, will likely have a substantial correlated response under another set of conditions, such as low-input farming.

Importance of Uniformity for Cultivar Development

Although new and alternative sources of germplasm may offer potential, it is unlikely that breeding, or most testing, is likely to be conducted under minimum-input conditions (Duvick, this volume, Chapter 4). For any substantial breeding progress to occur, breeding nurseries must be conducted under uniform conditions. Otherwise, genetic differences are masked by environmental differences (Souza et al., this volume, Chapter 10). One of the best ways of assuring environmental uniformity is to provide a surplus of nutrients, including fertilizers and water, and to eliminate pests using selective herbicides, fungicides, and insecticides. Successful plant breeding is quite dependent upon the establishment of uniform selection environments. If there is to be progress for disease resistance then a uniform threshold of disease needs to be established, quite often by inoculation. Similarly, if insect resistance is to be achieved, rearing and release of eggs or larvae is usually needed. Ideally, this is done in the absence of other diseases or insects to avoid confounding results and slowing progress from selection (Hoffman et al., this volume, Chapter 6).

The same concepts hold, perhaps even more strongly, for small-plot yield trials. When such conditions are not met, then selection merely implies choosing those plots that encountered fewer pests, flowered at the time when moisture was optimal, were located in the most favorable sites in the field, and so on. These are some reasons that so little progress has been made for drought tolerance; it is *very* difficult to establish a uniformly dry, uniformly fertile field that still supports plant growth. Similarly, developing cultivars resistant to sporadically and spottily occurring aflatoxins, well-known human carcinogens, awaits better methods of inducing uniform exposure (Payne, 1987).

Virtually all cultivar selection depends upon establishing uniform stands, which in turn depend upon planting seed of the highest quality. Thus, experimental lots of many types of seeds must be treated and handled with far more care (and usually far more inputs) than commercial lots of seed. Indeed, the quality of experimental seed lots should much exceed that of lots producing seed for routine planting by farmers. This has important implications for the specific uses of pesticides to protect experimental lots of seeds, because pesticide clearances are almost universally based upon large-scale commercial production.

Elite vs. Nonelite Germplasm in Cultivar Development
Remembering that the introduction of nonelite new germplasm is likely to lower yields and certain aspects of adaptability, how does one select germplasm for a breeding program that must produce cultivars for traditional and alternative farmers? First, the future success of any breeding program is likely to be today's best cultivars. Today's best cultivars are generally descendants of crosses made ten years ago between what were then the best cultivars. The exceptions to this rule occur when the germplasm base becomes too narrow, as is threatened in several crops, so that crosses solely among elite cultivars cannot provide the necessary response to changing environments, which can range from global warming to an increase in no-till cultivation. Under such conditions it would be prudent to seek additional sources of genetic variation among the most elite sources that promise to provide the needed trait(s). For example, germination under no-till cultivation may require early, low-temperature vigor, which could be sought from primitive Andean or Himalayan cultivars or from advanced cultivars from

37

northern Europe or northern China. All might be helpful, but the European or better Chinese cultivars might be helpful in half the time, because they probably have fewer detrimental traits.

Similarly, response to lower fertility levels might be sought among maize grown by farmers practicing slash-and-burn agriculture in tropical rain forests where soils are traditionally low in nitrogen. Progress would likely be more certain, however, and certainly quicker, if elite Argentine hybrids were used as source materials. Argentine farmers use little supplemental nitrogen yet produce yields comparable to those in the United States. Thus at least some Argentine germplasm ought to provide adequate genetic variability to cope with lower nitrogen use in the United States, and it is far more adapted to the United States than germplasm accessions from the tropics.

On the other hand, if genetic variation for insect or disease resistance is needed, then elite tropical cultivars are a likely source since planting, harvesting, diseases, and insects are continuous, year-round phenomena in many tropical areas of the world. Thus, nonresistant cultivars are unlikely to thrive there for long periods of time.

Sources of Genetic Variation for Yield

Often a regionally successful breeding organization may specialize in distinctive germplasm. Therefore, breeders seeking genetic variation for a particular trait, yield for example, are likely to find it in adapted populations developed by organizations specializing in distinctive germplasm. Sometimes elite foreign germplasm may provide the necessary variation.

In maize, where virtually all temperate breeding organizations worldwide use at least one parent from the same source population and many use the same single heterotic combination, it is a virtual necessity to exploit tropical sources to gain truly new variation for yield. This may include using breeding materials from those organizations that have used tropical germplasm in their own breeding programs. Thus, private companies such as Agroceres (Jacarezinho, Paraná, Brazil), Cargill (Minneapolis, MN), DeKalb Plant Genetics (DeKalb, IL), Morgan Seeds (Colón, Buenos Aires, Argentina) and Pioneer Hi-Bred International (Johnston, IA) serve as sources of breeding materials for other breeding programs. Public agencies such as CIMMYT (Mexico City, Mexico), IITA (International Institute for Tropical Agriculture, Ibadan, Nigeria), or those of the governments of Brazil, Colombia, Mexico, Peru, and Thailand that have had active breeding programs using tropical materials, also serve as sources of breeding materials.

Elite × elite crosses eventually can reach a point of no further genetic returns. It is unlikely that this has yet happened for any U.S. crop with respect to yield-potential under conditions of minimum inputs. Minimum-input farming is a relative newcomer to the U.S. plant breeding scene, so little selection for adaptation to low inputs has been conducted; hence, relatively little genetic variation for such adaptation is likely to have been lost. The daunting task is the evaluation of elite materials for new and difficult testing environments.

Germplasm, Economics, and Ecology

The evaluation of elite, domestic germplasm of major crops for use with restricted inputs can probably most readily be done by private companies through their extensive testing networks. Public institutions will likely be less able to react promptly, unless they have extensive and very uniform testing facilities. In many cases, years of tests will be needed to determine the stability and quality of germplasm performance under restricted-input cultivation. Thus, the initial germplasm that is likely to be used for low-input farming is the germplasm that has performed best under optimal inputs.

With foreign germplasm of major crops and for virtually all germplasm of minor crops, evaluation and initial breeding will most likely be done by public agencies. The return on investment is likely to be too low (Duvick, 1991) and the time frame too long (the time required from initial evaluation to final marketable product is often five times most private companies' "five-year plans"). The urgency of tomorrow is not apparent today—which emergency will be encountered a decade from now cannot be predicted with adequate economic certainty today.

Clearly, improved cultivars with higher pest resistance and more efficient nutrient utilization are the optimal solution to an economically sound agricultural system. Although the correct choice of germplasm is the key to a successful breeding program, successful breeding programs to provide appropriate cultivars must be the ultimate foundation for ecologically sound agriculture.

An Example of Incipient Change

Although dramatic changes in breeding approaches are unlikely in most major crops, they are far from impossible to implement. U.S. maize breeders are perhaps the most advanced and the most conservative group among today's major breeding fraternities. In maize, virtually all temperate breeding advances are achieved using elite × elite inbreds. Exotic cultivars are rarely used in the United States. Wild relatives are regarded as virtually useless. Tropical maize, which is highly sensitive to daylength, is effectively isolated from the temperate breeding populations generated by most maize breeders. When tropical materials are used they constitute very small percentages of final cultivars (Cox et al., 1988).

Thus, it was with skepticism that we viewed the first topcross yield trial results involving an experimental line that eventually became NC296 (an inbred line of maize derived from a cross of two commercial, all-tropical hybrids; Holley and Goodman, 1988). NC296 is not an exceptional maize inbred. It has several less-than-desirable features that will probably prevent it from ever being widely used commercially. First, it is white, not yellow. It has a gametophyte factor that helps keep it white but which prevents it from being much used as a female (seed) parent. It has less-than-desirable resistance to stalk breakage and has poor root strength as well. On the positive side, it has good resistance to several important diseases (gray leaf spot and southern rust, for example). Most importantly, it has excellent yield potential in hybrids and is unrelated to any other U.S. inbred. The performance of NC296 and its precursor lines is illustrated in Table 3.1, where its relative rank among commercial checks and competitors is shown for seven years of yield trials. Clearly, hybrids with NC296 as a parent yield competitively.

The more remarkable story is that NC296 was developed from a very small breeding and genetics program (started in North Carolina State Uni-

Table 3.1. Topcross yield performance of NC296 relative to commercial checks and other experimental lines.

Generation	Year	Tester	Rank among Checks[†]	Rank among Entries	No. Reps[††]
F_6S_1	1983	B73	2 of 4	2 of 39	5
F_6S_2	1984	A632xB73	2 of 8	3 of 42	9
F_6S_2	1984	Mo17x(H95xH99[3])	5 of 8	9 of 42	9
F_6S_2	1985	A632xB73	3 of 8	4 of 42	9
F_6S_2	1985	Mo17x(H95xH99[3])	4 of 8	9 of 42	9
F_6S_4	1986	B73xMo17	1 of 10	1 of 90	9
F_6S_4	1987	B73xMo17	3 of 10	4 of 36	9
F_6S_4	1988	B73xMo17	1 of 8	1 of 30	9
F_6S_7	1990	B73xMo17	1 of 5	1 of 42	9

[†]Rank among Checks refers to how the topcross ranked if it were considered as an additional check. Checks used (and number of years) included were: B73xMo17(9); B73xMo44(1); DeKalb 689(4) and 789(5); McNair 508(1); Northrup King N8727(1); Pioneer 3055(4), 3165(9), 3233(3), 3358(5), 3369A(6), and 3389(5); and USS 9001(2).

[††]Three reps at 3 North Carolina locations, except 1983, which was a one-location screening test.

versity's department of statistics). The family of lines from which NC296 was selected was established as part of a simple genetics experiment lasting from 1976 through 1982, the purpose of which was not to develop lines but to test whether daylength insensitive materials could be derived from daylength sensitive × daylength sensitive crosses. In 1983, the program was shifted to the crop science department and expanded. The key NC296 family, however, was derived from a total of 16 plots, each with 30 or fewer plants. Even under a much expanded crop science breeding program, the cumulative total of plots used for the whole daylength-response experiment was less than 1300 (ranging in size from 5 to 30 plants each, averaging less than 15), and the total number of plots used for NC296 and its several cousins was only about 350. This was thus a minuscule program by either public or commercial breeding standards, for which several hundred or more plots are commonly devoted *each year* to every set of families derived from a single cross. Thus the 21 crosses represented by the complete original set of families (7x6/2) would ordinarily be represented by 4000 to 7000 plots per season in most serious plant breeding programs.

During the development of NC296, the first five or six years of crosses among plants within lines were not on the whole very encouraging in achieving either daylength insensitivity or agronomically promising lines, as a discerning reader might imply from the virtually steady decrease in plot count for the experiment (from 36 plots in 1977 to only 19 plots by the spring of 1982) (Table 3.2). Had this experiment been conducted in a non-academic setting (or by a nontenured academician), it would have undoubtedly ceased by 1981 (or sooner, had an administrator ever visited the field). Yet, in addition to NC296, a large-scale breeding effort to identify and develop several families of totally new and usable parental inbreds has resulted from this academic investigation. Thus far, the results obtained in a very competitive arena with a crop that has not recently proved tractable to widespread infusion of new sources of germplasm suggest that with patience it would be possible to alter the germplasm base of a major crop in a dramatic fashion. The example also suggests that such a change is unlikely to be engineered by any but the most farsighted commercial companies, as there is no likely short-term profit in the endeavor. NC296 is at least a generation removed from profitability.

This example does suggest that if the sustainability of agriculture demands a major change in germplasm base, then change is feasible without great losses of productivity. The challenge is knowing what changes will be required 25 years hence and how to identify the appropriate parents to achieve the necessary changes. Our major limitations are evaluation procedures and lack of the long-term planning and support necessary to derive the ecologically compatible cultivars desirable in the future.

Table 3.2. Development of NC296, a temperate-adapted line derived from all-tropical germplasm.

| | | Number of Plots | |
Procedure	Year	NC296-Type Pedigrees	First Cycle Tropical Line Project
Initial F_1 crosses	1976	0	7
Grew F_1 hybrids	1977	2	36
Grew F_2 generation	1978	1	21
Grew F_3 generation	1979	1	20
Grew F_4 generation	1980	4	23
Grew F_5 generation	1981	3	21
Grew F_6 generation	1982	3	19
Grew F_6S_1 & S_2 generation	1983	46	159
Subtotals		60	306
Grew S_3's	1984	33	144
Grew S_3–S_5's	1985	76	399
Subtotals		109	543
Grew S_3–S_6's	1986	30	136
Grew S_3–S_7's	1987	36	115
Grew S_3–S_5's & S_8's	1988	2	15
Grew S_5's, S_6's, S_8's & S_9's	1989	69	82
Grew S_7's–S_9's	1990	35	35
Subtotals		172	383
Totals		341	1 232

References

Bauman, L.F. 1981. Review of methods used by breeders to develop superior corn inbreds. Proc. Annu. Corn Sorghum Ind. Res. Conf. 36:199–208.

Berlan, J.-P., and R. Lewontin. 1986. The political economy of hybrid corn. Monthly Review 38(3):35–47.

Castleberry, R.M., C.W. Crum, and C.F. Krull. 1984. Genetic yield improvements of U.S. maize cultivars under varying fertility and climatic requirements. Crop Sci. 24: 33–36.

Cox, T.S., J.P. Murphy, and M.M. Goodman. 1988. The contributions of exotic germplasm to American agriculture. p.114–144. In J.R. Kloppenburg, Jr. (ed.) Seeds and sovereignty: The use and control of plant genetic resources. Duke University Press, Durham, NC.

Duvick, D.N. 1977. Genetic rates of gain in hybrid maize yields during the past 40 years. Maydica 22:187–196.

Duvick, D.N. 1991. Industry and its role in plant diversity. Forum for Applied Research and Public Policy 6(3):90–94.

Fehr, W.R. 1987. Principles of Cultivar Development. Vol. 1. Macmillan, New York, NY.

Hallauer, A.R., and J. B. Miranda-F. 1981. Quantitative genetics in maize breeding. Iowa State Univ. Press, Ames, IA.

Holley, R.N. and M.M. Goodman. 1988. Yield potential of tropical hybrid maize derivatives. Crop Sci. 28:213–218.

McMillen, W. 1991. The ancient technology of farming: Ohio, 1910. Amer. Heritage of Invention and Technol. 7(1):44–49.

Meghji, M.R., J.W. Dudley, R.L. Lambert, and G.F. Sprague. 1984. Inbreeding depression, inbred and hybrid grain yield, and other traits of maize representing three eras. Crop Sci. 24:545–549.

Moll, R.H. 1978. Effects of recurrent selection for yield of corn. Proc. Annu. Corn Sorghum Ind. Res. Conf. 33:16–23.

Moreno-M., J.D. 1989. Temperate and tropical evaluations of temperate maize lines derived from totally photosensitive materials. Ph.D. diss., Dept. Crop Science, North Carolina State Univ., Raleigh, NC.

Payne, G.A. 1987. *Aspergillus flavus* infections of maize: Silks and kernels. p.119–129. *In* M.S. Zuber, E.B. Lillehoj, and B.L. Renfro (ed.) Aflatoxin in maize. Centro Internacional de Mejoramiento de Maiz y Trigo, Mexico, Mexico.

Russell, W.A. 1974. Comparative performance for maize hybrids representing different eras of corn breeding. Proc. Annu. Corn Sorghum Ind. Res. Conf. 29:81–101.

4

The Role of Seed Companies
in Crop Improvement

Donald N. Duvick

The first duty of a research-based seed company is to survive. The mode of survival is to breed, produce, and sell improved crop cultivars, and to do so at a profit. Profits depend on the difference between income from seed sales and expenses for operations and overhead (including expenditures for research and development).

Sales volume depends on the size of the customer base and also on the comparative performance and quality of the products offered by the individual seed company. Pricing of the seed products depends to some extent on comparative performance and quality. Pricing is also affected by the competitive price structure and by seed supply and demand.

Research and development therefore are critical to the success of a research-based seed company. Research and development may be done in-house, or research products may be purchased from other organizations such as foundation seed companies. Some seed companies depend almost entirely on finished products from the land-grant universities.

Sustainable agriculture will affect seed companies as it affects their sales, both in the nature of the products they sell and in the volume and price of their sales. Reciprocally, seed companies will affect sustainable agriculture as they provide or fail to provide seed products needed for the successful operation of sustainable agriculture.

Special Needs

Although sustainable agriculture is at present more a philosophy than a tightly defined set of procedures, it can be broadly defined as a system intended to reduce environmental contamination, conserve resources, and provide adequate and dependable farm income (Lockeretz, 1989). Sustainable agriculture should also "maintain the social fabric of the rural community" (Keeney, 1989). A further requirement, implicit although often unstated, is that sustainable agriculture will enable production of increasing supplies of food for a growing world population (Ruttan, 1988).

Some of the most commonly specified needs of such a system, with possible application to seed companies, are shown in the following list adapted from Lockeretz (1988, 1989), with additions and interpretations drawn from Jackson (1987), Goldberg et al. (1990), and Vereijken and Viaux (1990):

1. Diversity of crop species, to allow crop rotation and improve stability of income.

2. Crop cultivars with enhanced resistance/tolerance to insect pests and crop pathogens. These reduce the need for the purchase of chemical controls.

3. Crops with specific adaptation to relatively unmodified soil and climatic conditions. The goal is to work with soils and weather as they are found and to avoid the use of extensive purchased inputs to force different environments to a uniform standard.

4. Tillage systems that leave plant residues on the surface and crop cultivars that are adapted to the (often) new microclimate. Covered soil is colder and wetter in the spring and cooler and moister in the summer.

5. Preference for farm-generated or locally available inputs. As far as possible, the goal is to avoid the use of purchased materials, especially agrichemicals, or materials from large firms (nonlocal) of national or international nature.

6. Use of natural methods whenever possible for control of weeds, insect pests, and plant pathogens. Chemical controls are avoided in part because of

concern for environmental degradation and in part because of the desire to avoid purchased inputs.

7. Cover crops and green manures. These are intended to help prevent erosion and add nitrogen to the soil, thereby helping the environment and reducing the need for chemical purchases.

8. Rotations that include deep-rooted crops for more efficient mining of minerals from lower soil layers. This is intended to reduce the need for the purchase of commercial fertilizer.

9. Increased use of mechanical tillage and crop rotation to control weeds. This follows the general principle of using natural methods, or at least nonchemical methods.

Seed Companies and Special Needs

Seed companies may have products already on the market or nearly ready for release that meet some of the needs listed above. Comparative testing of cultivars now for sale, or in experimental stages, can identify any that meet the special needs of sustainable agriculture. On the other hand, if cultivars with needed adaptations are not on hand, seed companies may be able to breed them, drawing on their own or other available pools of genetically diverse breeding materials. In either case, new testing regimes may need to be devised and put in place to efficiently identify desired genotypes (Atlin and Frey, 1989).

In other cases, crops not now handled by the seed companies may be called for by adherents of sustainable agriculture. Seed companies might be able to expand their product line, breeding and selling these new crops.

BREEDING FOR PEST RESISTANCE

Probably one of the most important special needs of sustainable agriculture is for crops with high tolerance or resistance to insect and disease pests. Seed companies already place major emphasis on these traits and have done so from their earliest days. For example, large portions of most soybean breeding programs are devoted to the development of cultivars with phytophthora root rot resistance, and maize breeders devote significant amounts of effort to breeding maize hybrids with increased levels of tolerance to European corn borer.

But in many cases the tolerance/resistance levels of even the best cultivars are not high enough to give satisfactory control every year or in all locations, even when sustainable practices are used. Farmers will then need to use chemical control measures (when they are available) if they want to avoid unacceptably high levels of damage from pests.

For example, tolerance to European corn borer, second generation, is much better in modern maize hybrids than it was in the first hybrids of the 1930s. But in some years it still is not good enough. Farmers still lose yield and quality in their maize harvest in years when the corn borer is especially active.

In another example, the entire species of *Zea mays* seems to lack genes for tolerance or resistance to both northern and western corn rootworm. Standard breeding can do little to alleviate rootworm damage, typically severe in second-year maize fields. Without the use of insecticide, the second-year crop is liable to suffer from excessive root pruning, lodging, and loss of yield—all from the activity of corn rootworm larvae.

Biological agents that control crop pests are now being developed and in some cases are already available for use (Moffat, 1991). The agents may be applied separately, or they may be incorporated into the plant genome, enabling the plant to marshall its own defenses as needed.

It seems very likely that molecular genetics and genetic transformation will soon be able to add new resistance genes to the maize genome, genes with useful levels of resistance to corn borer and rootworm. But some of the practitioners of sustainable agriculture may not wish to use the products of biotechnology to enhance resistance to these insects (Jackson, 1987; Goldberg et al., 1990).

Three objections are raised, one based on scientific grounds and the other two on philosophical considerations.

First is concern that the imparted resistance would likely be the result of a single gene, which could soon be overcome by genetic changes in the pest population.

Second is a concern that even if they were designed to be effective and long lasting, such changes would be made by tampering with nature—by exerting excessive and unnatural control over nature rather than by working with nature—and as a consequence such changes would likely cause unexpected and undesirable consequences.

Third is a concern that even if both of the first problems were dealt with, the technology of making the change is so difficult and expensive that only the largest seed companies, or those owned by even larger agribusiness firms, could do the necessary research and development. Farmers would thus need to buy these inputs from large international firms, a practice some advocates of sustainable agriculture would prefer to avoid, as noted earlier.

MARKET SIZE AND PROFITABILITY

A more general restriction on the efforts of seed companies to breed for the special needs of sustainable agriculture has to do with market size and profitability. Potential markets must be above minimal levels in size and profitability if they are to attract research-based seed companies. The companies will not invest in the overhead costs of new research and development activities, or of new production and marketing activities, unless sales (and in the end, profits) from the new venture look to be high enough to warrant the investment. The present level of activity in sustainable agriculture is too low to attract special breeding and selection programs in the mainline commercial sector of plant breeding. This is true in regard either to the development of new crops or to breeding special new versions of presently sold crops.

Exceptions to the constraint of low market size and/or low profit margins might be in those cases where minimal additions to testing and selection procedures could result in new products with an adaptation to the unique needs of sustainable agriculture. For example, special attention could be paid to identifying maize hybrids with an unusually high degree of tolerance to the European corn borer, first generation. These could be recommended for use by farmers who need better protection from first generation borer but want to avoid the use of insecticides.

This example points up the most likely way that seed companies can assist sustainable agriculture. Their cultivars can be screened and labeled with regard to the traits especially needed for use in sustainable agriculture. In fact, many seed companies have trait ratings on their products at the present time. These ratings could be adapted for use in sustainable agriculture. If proponents of sustainable agriculture, as a body, could present a unified list of the most needed traits for each crop in a given growing region, the seed companies could probably identify the cultivars they now sell that come closest to meeting those needs. And with such lists in hand they also might be stimulated to expand their screening tests to cover points that were lacking.

PREADAPTED CULTIVARS

It will probably be found that many cultivars are already well adapted to many of the needs of sustainable agriculture, even though such adaptation was not a specific goal when breeders made their selections. For example, selection for geographically wide adaptation and stable performance over the years (standard breeding practice) leads to a choice of cultivars with improved performance in poor as well as fertile soils, dry as well as wet seasons, disease or insect epidemics as well as pest-free seasons, and hot as well as cool growing conditions. Such cultivars are preadapted to low-input farming and farming that places a premium on stability of performance.

Some cultivars are much more stable than others under such a wide range of conditions. Breeders now identify such cultivars, and these exceptionally stable products could be earmarked for use in sustainable agriculture, if such a trait were desired. Some highly stable cultivars may not be in the highest yielding group in highly favorable stress-free growing conditions, and farmers will need to decide, in such cases, whether they want gamblers' odds or insurance. In general, practitioners of sustainable agriculture might choose cultivars that give them insurance.

But some cultivars are highly stable and also give superior yields under highly favorable conditions. Such cultivars are appearing with greater frequency as the years go by, as plant breeders improve their breeding stocks and their breeding techniques (Duvick, 1992). These cultivars will be the first choice of all farmers.

Despite generally favorable prospects for finding crop cultivars well suited for sustainable agriculture, some desired traits will not be available in cultivars developed by commercial plant breeding. These will be new traits for which the breeding is so difficult and time-consuming that seed companies cannot afford the extra expense over a long time period, particularly if the potential market is small and poorly defined.

An example might be development of crops better adapted to organic than to chemical fertilizers. In principle, one might develop cultivars better able than those now on hand to cope with delayed release of nitrogen in the springtime (characteristic of organic sources) and to take advantage of any other subtle differences in timing and form of available mineral nutrients. But the difficulty of detecting differences, the complexity of test design, and the possibility of differences in action among different forms of organic manures all present such a difficult, expensive, and long-range challenge that commercial seed companies very likely will not attempt to breed or even to test for cultivars with better adaptation to organic manures, unless the demand for such cultivars grows to a significant proportion of the total market of seed buyers. Such research is more properly in the domain of state and federal public institutions.

Several other needs of sustainable agriculture are also not likely to be supplied by the seed companies. Commercial breeding will probably not be done for minor crops used only in sustainable agriculture, for crops adapted to unusual soil types, for crops to be used as cover crops and green manures, or for crops with extra deep, mineral-mining roots. The reasons listed above for not breeding for special adaptation to organic manures apply to this list of special needs also.

PROFITS

One further possible hindrance to actions of seed companies in aid of sustainable agriculture is the profit-oriented nature of the seed companies and its effect on pricing and ownership of germplasm, their stock in trade. As noted earlier, the seed companies depend on sales of proprietary products for their existence. Prices must be high enough to cover all expenses, including research and development, and to give a satisfactory margin of profit. Seed prices for privately bred cultivars therefore will be higher (other things being equal) than for cultivars bred with public funds and released without charge. The cost of private plant breeding is borne by the seed purchaser rather than by the general public.

This fact has long been recognized and both farmers and seed companies know that the commercial cultivars will need to give extra profit to the farmer if the extra price is to be justified. Since a usual goal of practitioners of sustainable agriculture, however, is to reduce input costs, the increased price of commercially bred seeds may deter purchase of them, unless it is seen clearly that to spend more on certain seed cultivars will increase farm profits in sustainable fashion. On-farm trials and careful analysis of their results will be needed for sound decisions on this matter.

Another objection to dealing with seed companies could be that they maintain proprietary rights over the products of their research. These rights can be in the form of trade secrets (the pedigrees of hybrid maize varieties, for example, are not revealed); plant variety protection certificates that forbid unauthorized parties to reproduce the cultivars for general sale; or patents on cultivars, plant parts, products, genes, or processes. All of these forms of proprietary control run counter to the general desire of those practicing alternative agriculture to use farm-generated or at least locally produced inputs.

A further complication is the close connection that many seed companies have with agricultural chemical companies, either because they are subsidiaries of such companies or because they have formed some kind of business liaison to facilitate development of cultivars with special products or adaptations that fit with the business plans of the chemical companies. Many of those espousing sustainable agriculture practices fear that such liaisons will result in chemical-dependent crop cultivars or at least genetically engineered crop cultivars—products of biotechnology. As already noted, such products are not well regarded by some proponents of sustainable agriculture (Goldberg et al., 1990; Jackson, 1987).

How Seed Companies Can Help

Despite this list of possible problems there is a longer and weightier list of ways in which research-based seed companies can help sustainable agriculture to reach its goals.

COOPERATION WITH GROWERS

Seed companies, as already noted, select under a multitude of environments, including many of those needed for the best practices in sustainable agriculture. For example, they devote much attention to selection for disease and insect resistance, one of the most important traits needed in crops suited for sustainable agriculture. With clearly defined lists of needs, sustainable agriculture practitioners working with seed companies should be able to select cultivars that most nearly suit their needs.

ON-FARM TEST SITES

When (or if) the proportion of acres devoted to clearly recognizable sustainable practices reaches significant size, seed company breeders will automatically include sustainable farms in their list of standard on-farm test sites. The history of plant breeding, commercial and public alike, is that as new farming systems develop, testing and selection are done using the new systems in order to develop cultivars with adaptation to them. The most successful seed companies are those that best forecast important farming system/cultivar interactions for the future and then are the first to develop superior cultivars to fit those new needs. But correct forecasting and proper timing of research and development activity are critical to success in these endeavors.

DIVERSITY

Seed companies are the prime developers of diverse new kinds of useful germplasm in the form of finished cultivars. They promote genetic diversity, an important goal of sustainable agriculture. This statement may seem counter to the general impression that seed companies narrow the germplasm base of farm crops. In actuality, the demand from farmers for only the top performing cultivars and the demand of the marketplace for uniformity in crop products are the primary factors in narrowing the base.

Seed companies constantly struggle to broaden the variety of their offerings, in part because this helps in their continuing struggle to capture market share from competing companies and in part because they know that since no cultivar is perfect, they as seed companies are protected by having a genetically diverse set of cultivars for sale at all times. Genetic vulnerability for a seed company is to have most of its sales volume concentrated in a few closely related cultivars.

Further, the diversity of cultivars on the farm in any one year represents only the tip of the iceberg. Much more diversity is found in experimental cultivars under trial by seed company plant breeders, and much more yet is found in their germplasm pools, from which new cultivars are selected. All of these materials could be used with little delay, should the need for them arise.

In addition, the rapid replacement of cultivars by improved new cultivars—seven years on the market is an average lifetime for a cultivar once it is released—means that genetic diversity in time is operating in favor of the farmer. Pest biotypes have only a few years for selection and adaptation to any particular cultivar.

Also, different parts of the country require different cultivars; those adapted to the northern states are not suited for southern states, for example. Thus, a patchwork pattern of genetic diversity always exists, countrywide, in any one growing season.

Despite these actions of commercial breeders in promoting diversity, it must be said that public plant breeding and genetics have an extremely important and fundamental role in fostering genetic diversity (Goodman, this volume, Chapter 3). Public breeding has the assignment of maintaining and developing even broader pools of genetically diverse breeding materials. Public breeding programs can work on longer time scales and take greater risks with untried new germplasm than is possible for most seed companies, and the mandate of public plant breeding programs is to discover, develop, and refine such basic new breeding materials. They can then be added to proprietary elite breeding pools, from which finished cultivars are developed and delivered to the farm by seed company breeders.

Seed companies understand the vital role of public research in plant genetic resources. They know it is important to farmers and national and international welfare, as well as to their business. In recent years seed company scientists have taken active roles in crop advisory committees of the National Plant Germplasm System. Through trade organizations and membership on public-private advisory panels, they and other seed company officials have urged increased federal and state investment in plant genetic resource conservation and use. Some companies also have donated their own funds to this purpose, in the form of special unrestricted grants and fellowships.

UNBIASED INFORMATION

Seed companies often employ agronomists to advise farmers on the best practices for growing the company's cultivars. The seed company agronomists are not beholden to chemical companies, to machinery manufacturers, or to public extension services. Therefore, they are usually regarded by farmers as reliable and relatively unbiased sources of information regarding those nonseed products (there may be some concern if the seed company is owned by an agrichemical corporation). As sustainable agriculture practices become more clear-cut and as they prove their worth, seed company agronomists will promote and explain them, knowing that such knowledge can help farmers to get the best performance from their companies' cultivars. New tillage systems, for example, are already included in the demonstration and experimental plots of the seed company agronomists. And when and if certain cultivars are found to be especially well adapted to widely used sustainable agriculture practices, they will be promoted by the seed companies through use in agronomic demonstration plots and in on-farm comparison trials.

INNOVATION BY SMALL SEED COMPANIES

A source of continuing innovation and strength in the seed business is the multitude of small seed companies. Up to 40% of U.S. hybrid seed maize is sold by small companies, for example. These companies provide useful competition to the larger companies (at least from the farmer's point of view), since they are usually willing to sell seed at lower prices; they are very likely to spring up in areas that have few competitors; and they are continually trying out new ideas, new kinds of products, and projects that the larger companies may be unwilling or slow to undertake. In particular, they are good at filling niche markets, special small markets that can be overlooked by the larger companies.

Such small seed companies may be the ones that first dedicate themselves to serving any unique needs of sustainable agriculture systems. They might select and sell seed of new kinds of crops, such as interplant legumes for maize. They might search out, from the offerings of public researchers, alfalfa cultivars with especially high rates of nitrogen fixation and multiply them for sale to interested purchasers. They might develop, or more likely get from public researchers, wheat cultivars with special adaptation to acid soils and sell them to the restricted market of potential purchasers.

Practitioners of sustainable agriculture might well look to small seed companies for initial interest and action in filling their special needs. They might even consider starting their own cooperative plant breeding companies or sponsoring a small company focused on their needs. Such companies might have contractual relationships with one or more large companies to supply certain technology or germplasm that they could not develop on their own. Thus the benefits of small size and large size could be combined.

Conclusions

Seed companies can fill many of the needs for seed cultivars well adapted to sustainable agriculture practices. Breeding for wide adaptation has resulted in the development of many cultivars that once identified will fit well with sustainable agriculture practices. Seed company agronomists can also help in testing and demonstrating cultivars for adaptation to sustainable practices, as well as demonstrating sustainable practices as such. But the small market presented by sustainable agriculture at the present time, the lack of general agreement on tenets of sustainable agriculture, and the difficulty of breeding for some special traits will probably prohibit commercial activity in at least some of the special seed needs of sustainable agriculture. Small entrepreneurial companies may be useful pioneers in breeding or procuring seeds with special traits needed for sustainable agriculture. A matter of particular interest is the reluctance of some in sustainable agriculture to deal with products of genetic engineering, with patented or otherwise protected seed products, or with large international firms connected to agrichemical companies. In the years to come, most of the nation's commercial plant breeding will involve all of these factors.

References

Atlin, G.N., and K.J. Frey. 1989. Breeding crop varieties for low-input agriculture. Am. J. Altern. Agric. 4(2):53–58.

Duvick, D.N. 1992. Genetic contributions to advances in yield of U.S. maize. Maydica 37:69–75.

Goldberg, R., J. Rissler, H. Shand, and C. Hassebrook. 1990. Biotechnology's bitter harvest: Herbicide-tolerant crops and the threat to sustainable agriculture. The Biotechnology Working Group, Environmental Defense Fund, New York, NY.

Jackson, W. 1987. Altars of unhewn stone. North Point Press, San Francisco, CA.

Keeney, D.R. 1989. Toward a sustainable agriculture: Need for clarification of concepts and terminology. Am. J. Altern. Agric. 4(3,4):101–105.

Lockeretz, W. 1988. Open questions in sustainable agriculture. Am. J. Altern. Agric. 3(4):174–181.

Lockeretz, W. 1989. Defining a sustainable future: Basic issues in agriculture. Northwest Report 8 (Dec.):1–13.

Moffat, A.S. 1991. Research on biological pest control moves ahead. Science 252:211–212.

Ruttan, V. 1988. Sustainability is not enough. Am. J. Altern. Agric. 3:128–130.

Vereijken, P., and P. Viaux. 1990. Vers une agriculture "integree." La Recherche Agronomie, supplement La Recherche 227:20–25.

5

Crop Breeding Objectives and Methods

Charles A. Francis

There is an accelerating move toward crop production systems that are more resource efficient, environmentally benign, yet profitable in both the short and long terms. Most farmers and research specialists agree in principle on the importance of improving overall soil productivity and potential, finding alternative fertility and pest management strategies, and reducing purchased input costs if this can be done without sacrificing profits. Many are moving in this direction, although some reject the term *sustainable agriculture*. Broad reviews of the principles and practices of future farming systems have been published by Edwards et al. (1990) and Francis et al. (1990).

Are new and specific cultivars needed for sustainable agricultural systems? This depends on the nature of these systems and whether current cultivars will perform well in them. What will be the growing conditions faced by plants in tomorrow's systems? To answer this we need to make assumptions about the cost and availability of inputs and put a time frame on sustainability. Are existing cultivars adapted to a wide enough range of systems to provide the genetic potential for maximum yield or profit under new sets of conditions? It is important to envision future systems and begin breeding for them now if such systems appear to require a drastic change in cultivars. These are the first and most important questions to address in a breeding program designed to improve crops for the next century.

Sustainable systems have been variously characterized as operating with

59

lower input use, making more efficient use of scarce resources, and providing more stable production with less risk. Coffman and Bates (this volume, Chapter 2) conclude that breeding for sustainability is a process of fitting appropriate crops and cultivars to an appropriate environment instead of altering the environment through inputs such as fertilizer, water, and pesticides. Their thesis coincides with the growing consensus that future systems will be more biologically structured to take advantage of soils, biota, and climate as we learn more about natural and modified ecosystems (Francis et al., 1986). Designing farming systems that build on the complex biological interactions found in natural systems provides a stark contrast to current conventional systems that attempt to dominate the natural environment with massive inputs of fossil fuel–based products. Most current production strategies have little regard for long-term resource scarcity or environmental impact. Some innovative scientists are exploring new options. An articulate description of the potential of perennial prairie cropping systems that build on natural biological integrities is made by Jackson (1980) and presented in more detail by Soule and Piper (1992). According to Coffman and Smith (1991) a sustainable agriculture is defined as "one that provides for the replacement of any resources removed (on a global scale)."

Most evidence suggests that future systems less dependent on fossil fuels will be characterized by higher levels of stress on plants. Plant breeders have been relatively successful in providing genetic tolerance or resistance to some types of stress (such as pathogen or insect resistance) and less effective in finding productive cultivars that will withstand rigors of climate (such as tolerance to drought or waterlogging). Differences in breeding success generally correspond to heritability and the complexity by which traits are controlled. There have been excellent reviews of genetic improvement of crops for stress tolerance (Blum, 1988; Christiansen and Lewis, 1982). Crop improvement for sustainable systems defined in the current context has been reviewed by Francis (1990) and summarized by Sleper et al. (1991).

To explore the objectives and methods in breeding crops for sustainable systems, this chapter defines a process for setting breeding objectives based on the best possible projections for future systems and accompanying stress conditions. Choice of parent material and decisions on specific breeding strategies are explored. Locations and conditions for testing progeny and the use of this information for making decisions on the best future cultivars are key determinants of success in a breeding program. How genetic improvement interfaces with changes in other component technology and whether or not breeding is the best solution to solve specific constraints are important considerations. Finally, the success of plant breeders in contributing to future system productivity under any set of conditions depends on how well those breeders can anticipate the changes that will take place in the cropping system environment. Guidelines for designing future cropping systems are explored elsewhere (Francis and Callaway, this volume, Chapter 1).

Setting Breeding Objectives

To choose appropriate parents, breeding and selection methods, and testing environments, it is critical to first decide on precise objectives. Standard texts place far more emphasis on the details of crossing and selection methods than on the choice of program objectives. Graduate students who spend two to five years in apprenticeship with a plant breeder in the university may miss this step all together, since objectives are most likely in place and a program already moving in a certain direction when they arrive. It is important to keep breeding objectives in mind and continually assess their relevance.

An implicit goal is the improvement of yield potential in a given crop. The strategies to achieve that goal include the introduction of yield-enhancing genes from outside sources, continued selection for yield, improved tolerance to biotic or climatic stress, and adaptation to specific sets of farm circumstances. These are all useful approaches, if they address current constraints to productivity that will likely continue into the future. Too often our goals are short-sighted, set in response to an immediate need or perceived challenge that may or may not be important in the long term. A one-year drought that has a probability of occurring once every 50 years should not precipitate a massive shift in a breeding program to selection for drought tolerance. On the other hand, the impressive response by plant breeders to the challenge of southern corn leaf blight that attacked maize carrying the t-cytoplasm sterility factor provided an appropriate and economically valuable contribution to maize producers in this country. It is essential to put these production challenges and the potential plant breeding response into perspective, both in terms of geographical relevance and long-term importance.

Suggestions for developing more permanent or durable resistance to pests have been proposed (Alexander and Bramel-Cox, 1991). These include breeding cultivars that have inherently durable resistance, modifying the agroecosystem to maintain resistance in cultivars, and developing cultivars that provide high yields in spite of infestation or infection. These all involve putting less pressure on pest populations to change and become more virulent.

As energy costs drive up the price of nitrogen and research reveals more details about how to manage nutrient cycles and biological sources of nitrogen, there will be greater interest in the current efforts to increase nitrogen-use efficiency of crop cultivars. As land becomes more scarce, it will become increasingly important to adapt crops to more marginal soil nutrient situations. The selection of crops for problem soils has been reviewed by Devine (1982).

An important question is whether or not new cultivars or hybrids will be needed for "sustainable systems" as compared to "conventional systems." Do the cultivars under development today have the adaptation needed for tomorrow's prevalent systems? It is difficult to answer this question without a precise definition of the conditions of future systems and without having conducted a comprehensive selection program for those conditions (Francis, 1990). Goodman (this volume, Chapter 3) suggests that current cultivars not selected for low-input conditions are not likely to be optimum for those systems. A good indication of the adaptability of cultivars, using available materials, is a study of genotype by system interactions (Souza et al., this volume, Chapter 10). If there is a significant genotype by stress condition interaction or a consistent genotype by tillage or a genotype by location interaction, it is likely that genetic progress can be made by selecting for the specific conditions that are likely to be found in the new systems.

What will characterize these future cropping systems? It is likely that the emphasis will be placed on maintaining yields under a range of climatic, soil, and biotic stress conditions; reducing input costs; and breeding cultivars adapted to specific systems and agroecological niches on farms. Although it is desirable from the perspective of a plant breeder or commercial seed company to breed single cultivars with a wide range of adaptation, it is likely that the genetic potential for specific adaptation will become relatively more important as inputs are withdrawn and systems become more diverse and unique to the conditions of a given field or farm. Some of the specific traits that are likely to be needed in selection for sustainable cropping systems are summarized in Table 5.1.

There are varied probabilities of success for these several traits. Devine (1982) suggests four requisites for success in genetic selection for traits associated with these lower input systems: (1) heritability of the trait, (2) availability of efficient techniques for identifying and screening for the trait, (3) desirable genetic variation in parental material (but see McCouch et al., this volume, Chapter 9), and (4) the potential for genetic improvement versus other solutions make breeding a logical approach to solving that constraint. Most of these requisites can be evaluated from past literature and practical breeding experience with similar crops and situations. For example, if breeding for greenbug resistance in sorghum has been successful, it is likely that breeding for resistance to a related insect in another crop will be successful. Selection for drought tolerance, on the other hand, has been frustrating for plant breeders because of the complexity of inheritance, the inconsistency of timing and severity of drought, and the difficulty of providing relevant testing conditions. Thus, the probability of success in a breeding program can be predicted with some degree of confidence.

In summary, it is important to set explicit goals, to make them realistic in terms of what is possible through genetic improvement, and to assess progress and readjust goals accordingly. Of prime importance is deciding which systems are most likely to prevail a decade hence and evaluating what stress

Table 5.1 Specific adaptation traits that will be needed
for sustainable systems.

Trait	Probability of Success	Reference
Insect resistance	High for most species	Hoffman et al., this volume, Chapter 6; Jenkins, 1982
Pathogen resistance	High for most species	Bell, 1982; Hoffman et al., this volume, Chapter 6
Nematode resistance	Moderate to low	Sasser, 1982
Competition tolerance	Variable	Callaway and Forcella, this volume, Chapter 7; Francis, 1986; Liebman and Janke, 1990
Drought tolerance	Moderate	Blum, 1988
Problem soil tolerance	Moderate	Blum, 1988
Nitrogen-use efficiency	Moderate	Zweifel et al., 1987
Phosphorous-use efficiency	Moderate	Clark et al., 1978
Root system modification	Moderate	Zobel, 1975

conditions are most likely to limit crop production in those systems. The process of goal setting should be an on-going activity, both to fine-tune the direction of a breeding program and to fully involve each generation of graduate students in this vital activity. It is highly unlikely that the best genetic materials and most rigorous breeding methods will provide success if the goals are not appropriate to the crop and systems into which new cultivars will be introduced. Explicit goals are the first essential step to the success of a plant breeding program.

Choosing Parents and Breeding Strategies

It is important to begin any breeding program with the best available cultivars as parent materials, consistent with the goals set for the program and the traits to be combined. Although the incorporation of exotic germplasm is sometimes needed when traits of interest are not otherwise available, this is a somewhat drastic step in that it introduces a number of other genes that will probably detract from agronomic performance, at least in the early stages of selection (Goodman, this volume, Chapter 3). Fehr (1987a) suggests the following sources of parental germplasm listed in their order of similarity to local commercially grown cultivars: other commercial cultivars, elite breeding lines, acceptable breeding lines, plant introductions of the same species, and cultivars from related species. The better the yield potential of the donor parent of a given trait, the more likely the progenies are to have high yield potential. Goodman (this volume, Chapter 3) suggests that elite lines or previously discarded materials exist in some other part of the world for most of the widely grown crop species and that this genetic base could be invaluable for future breeding programs. As plant breeders move on to other objectives or as pathogens and insects change, elite lines are put aside or maintained in nurseries or seed storage. These are often available to other breeders on a request or exchange basis. Although patents and plant protection legislation have slowed germplasm exchange in recent years, there is still a healthy communication and interchange of materials among most breeders.

It is not always possible to locate genes for tolerance to a specific insect, pathogen, or stress condition in adapted materials; it is then necessary to use exotic lines, introductions, or other species in place of the preferred elite materials. The consequences of such a choice include the need for larger numbers of progeny to find acceptable combinations, loss of yield potential, and the need for more cycles of selection in the chosen breeding method. The choice of exotic materials is not one to be taken lightly. Goodman (this volume, Chapter 3) concludes that although today's elite materials may not have been selected under conditions of minimum inputs, they still have much better adaptation, stress tolerance, and yield potential than their progenitors.

65

The choice of parents often includes highly adapted and productive cultivars or inbred lines for one side of the cross and an introduced cultivar or line with the trait or traits of interest on the other side. Locating potential parents for the trait of interest is generally successful by screening under the insect, disease, or stress conditions that characterize the new system of interest. Exceptions are traits under complex genetic control or traits that depend on heterosis found only in first-generation hybrids. In these cases, it is necessary to cross a large number of potential new parents onto the highly adapted parent and test the progeny. This lengthens the breeding process but is critical for future success. Sources of parent materials are the small germplasm collections maintained by most plant breeders, nursery collections of other breeders in cooperating programs in neighboring states or countries, or the national and international germplasm banks maintained by individual nations and through the Food and Agriculture Organization and the international agricultural research centers. These institutions operate free of political affiliations or alliances and make genetic material available to any plant breeder who makes a request. As Jensen (1980) points out, the choice of parents is a critical step in the breeding process, because this in large part determines the ultimate success of a program some years later.

There is evidence that known variation in response to current conventional systems may not be all the variation available to improve crops for distinct systems. For example, Atlin and Frey (1990) concluded from their studies of oat in different environments that different sets of alleles may be needed to maximize genetic performance in low-fertility versus high-fertility sites. This is especially critical in crops such as small grains, where soil fertility contributes directly to protein percentage and where this is one factor in the pricing strategy and market value. Where selection and testing should take place depends on the heritability of the trait of interest (that is, yield) in the target environment compared to the selection (or testing) environment, as well as the genetic correlation between expression of that trait in the two environments (Allen et al., 1978; Rosielle and Hamblin, 1981). More research is needed in this area of breeding methodology.

66

It is generally concluded by plant breeders that existing breeding strategies are adequate for improving crops for sustainable systems. Coffman and Bates (this volume, Chapter 2), however, suggest that the complex traits needed for stress tolerance or resistance are both difficult and expensive to select, and for this reason there is a need for more efficient breeding or screening techniques. It is likely that some of the genetic engineering techniques now being applied to crop cultivars will benefit breeding for sustainable cropping systems (McCouch et al., this volume, Chapter 9). The time-honored selection methods that have proved successful in the past with economically important species are most likely the methods that will persist in future crop improvement programs. These methods are amply reviewed in recent plant breeding texts and references (Blum, 1988; Fehr, 1987a,b; Jensen, 1988; Kalloo, 1988a,b,c).

Locations and Conditions for Testing Progeny

When plant breeding programs are focused on a relatively narrow range of production conditions controlled by irrigation, pest protection, and high levels of applied fertilizer, it is possible to carry progeny for a number of generations under similar conditions until sufficient seed of highly refined genetic combinations are ready for semicommercial testing. To breed crops for a wider array of unpredictable conditions, such as those found in sustainable systems, it is essential that progeny testing begin as early as possible so that as many genetic combinations as possible that do not contribute to the specific goals of the project may be eliminated.

The experience and procedures in early testing are discussed in some detail in Fehr (1987a), including specific applications for self- and for cross-pollinated crops. In general, it is possible to test in the F_2 generation (second generation after a cross) of self-pollinated crops if there is sufficient seed and if traits are simply inherited and can be identified in the progeny (this is known as the F_2-derived or F_2 family method). In cross-pollinated crops, the first generation of selfed progeny (S_1) is crossed to a tester while simultaneously selfing all lines of interest. Results of the testcross trials are used to choose the lines that will be kept in a program. In either case, it is necessary to advance new combinations to the stage at which they have enough genetic identity to react to the range of environmental or cultural conditions of interest and have a heritability high enough to maintain that desired reaction. The constraint of sufficient seed for testing in several environments is another limitation to early generation testing. Advantages of early testing include the early discard of inferior individuals, lines, or populations and the potential selection of more than one cultivar from a given line or population since these are still heterogeneous in early stages of selection. Disadvantages include resource diversion limiting the testing of more highly inbred lines and the potential delay in releasing a new cultivar if all materials cannot be tested and advanced simultaneously (Fehr, 1987a). If testing in multiple sites or conditions is important, the quantity of seed may prove too limiting in early generations to make this feasible.

The number of sites needed for testing depends on the range of conditions to be encountered by new cultivars and the stability of performance across those sites. Guidelines for choice and characteristics of sites are described by Bramel-Cox et al. (1991). They discuss the influence of environments on heritability and the advantages of wide versus narrow adaptation. Stability of performance is generally measured by analyzing the genotype by environment interaction. Souza et al. (this volume, Chapter 10) distinguish among five types of stability (biological, agronomic, linear, temporal, and across fixed environmental effects): "It is virtually impossible to develop a specific cultivar for each crop rotation or intercrop combination . . . ," thus it will be important to develop "genotypes that are reliably productive across a range of management schemes . . . if true cropping diversity tailored to local environments is ever to be achieved." Specific traits that influence performance and stability are described by Smith and Zobel (1991), including canopy structure, crop maturity and early development, and tolerance to stress and biotic factors.

Although choice of crossing and progeny testing methodology is not necessarily unique to breeding for sustainable systems, decisions on locations and management conditions are critical to breeding program success. Whether improved cultivars that come out of a breeding program will make

68

an impact on productivity depends on how closely the breeder is able to anticipate and then duplicate the conditions and constraints of the cropping environment in which new materials will be used. Souza et al. (this volume, Chapter 10) suggest, "the environments should be as diverse and adverse as the conditions for the target growers of the improved genotypes." Callaway and Forcella (this volume, Chapter 7) describe choice of environments as depending on how widely applicable selection environments are for future growing environments and what the ease or cost of developing cultivars for those future environments is.

The number of testing environments or sites depends on what seed supply is available, how early in the development process testing is started, and what resources are to be dedicated to this part of the program. Jensen (1988) presents a detailed discussion on choice of environments and sites for cultivar testing. Producer acceptance of a new cultivar depends in part on yield stability and responsiveness to favorable growing conditions. For individual farmers, yield potential for their specific fields over a series of conditions likely to occur over time is far more important than broad adaptation over sites. Yet sites can in some degree substitute for years, if the conditions encountered across those sites (especially rainfall) represent the array of possible rainfall years that will occur at one of the sites.

Producers have become more interested in stability, leading breeders to use more locations and less replication per location each year (Bradley et al., 1988). The potential for single replication testing with multiple locations has been studied by Dofing and Francis (1990), who conclude that optimum allocation of resources depends on the relative cost of replication versus the cost per location. There is also a question of how to use the information collected from multiple locations. Should all data be given equal weight, or should some sites that represent a wider area of use be given greater weight? Statistical weighting methods, one form of index selection (Baker, 1986), have been developed, but the majority of breeders continue to give equal weight to data collected from multiple sites. Ultimate success in any breeding program will depend on the degree to which new cultivars are multiplied and adopted by farmers. Their genetic value in

69

farmers' cropping systems will largely be a function of the testing program used for their evaluation.

Extensive Testing and Release of Cultivars

Advanced testing of promising cultivars is an extension of the process of progeny evaluation described above. Differences include the availability of greater quantities of seed, fewer cultivars (since many have now been eliminated), more testing sites, and larger plot sizes. Since these potential new cultivars are nearing release to the public, there is greater reason to test new genetic combinations in current cropping systems and under current constraints. In general, universities or other public breeding programs tend to conduct more years of tests with more replications in fewer locations, while commercial companies tend to use more locations and fewer years and replications. Perhaps this is because of the urgency placed on rapid release by companies and the competitive nature of the business. In either case, it is important to include the cropping systems and locations of most interest to farmers who will make the ultimate decision to adopt new cultivars.

One approach used by state experiment stations is the uniform hybrid or variety test. Current commercial and some experimental cultivars are submitted by companies to state university programs for testing. A minimal fee is charged for testing cultivars to partially cover the cost of these uniform trials. Several locations, both on-station and on-farm, are used to test these materials in a standard procedure. Many companies choose to include a number of their best cultivars and one or more experimental cultivars in these tests, since this provides an objective comparison and some publicity surrounding the results of the tests. Results of the uniform tests are published by the university programs and distributed.

An approach employed by companies to maintain the high popularity of their cultivars with farmers is the use of highly visible test strips. In these demonstrations, a number of current cultivars and promising lines from a given company are planted in plots perpendicular to highways so that many people will see them all season long. In collaboration with farmers, who are most often dealers for that company, these plots are well tended during the season and signs are placed so that each cultivar is identified. In the past these were used only as demonstration plots. Refinements in design and analysis, however, including the use of periodic check plots of a single cultivar for comparison and generating an error term, permit the combination of harvests from strips across locations using each site as a replication. In such a program, a company is able to provide data on each cultivar across dozens if not hundreds of sites in a region. This activity has become both practical research and demonstration. If these plots include the sites of greatest interest to farmers who are pursuing a sustainable management approach, the results can be useful. If demonstration plots are located only on the most productive soils and with high-input management (for example, sites where the cultivars will look exceedingly good), the results will be of limited value.

Release of new cultivars, either by public agencies or private companies, is discussed by Fehr (1987a) and in other plant breeding textbooks. There is probably little that is unique about the release process that merits note in a discussion of breeding for sustainable systems, although there is a segment of the population interested in alternative agricultural systems who insist that farmers become independent of the commercial establishment in their production and use of seed. Since hybrid seed must be reconstituted each year, this is seen as an outside production input that must be purchased and thus does not contribute to the sustainability of the system. Likewise, it could be argued that seed of self-pollinated crops such as soybeans or small grains can be saved from one season to the next by the farmer, thus making the genetic part of the system more "sustainable." There is nothing magic about the production of quality seed. Any serious producer can learn the techniques of combining inbred lines to produce hybrids of cross-pollinated crops, or selecting and producing quality seed of self-pollinated crops. Seed production is a specialized industry, however, and much care must be taken to preserve the genetic integrity of cultivars. It is also important to process and store the seed in a place and manner that assures quality seed for the next season. Germination tests are essential for determining proper planting rates. Older cultivars may have less genetic potential than those more recently released. For example, maize breeders are making at least one bushel per acre in genetic progress per year, while soybean breeders are increasing genetic potential by at least one-third bushel per acre per year. This assumes that genetic progress in current breeding programs provides genetic gain that is useful in more sustainable systems. In fact, most producers who explore the complexities of quality seed production decide that this is a difficult and time-consuming process and that purchase of seed is one of the most cost-effective decisions made in any production system, "sustainable" or not.

In summary, seed increase, final testing, and cultivar release for sustainable cropping systems are similar to those for conventional systems. Careful choice of testing environments is a key element to success. Seed production on the farm must be carefully evaluated as an option to diversify the farm enterprises, assure a source of quality seed, and perhaps reduce purchased input costs. Most farmers find other ways to reduce inputs that are more cost-effective in terms of their investment of time and facilities.

Program Evaluation and Long-Term Planning

Evaluating success in a plant improvement program is a continuing process with one key criterion, the amount of seed of each new cultivar distributed or sold. The degree to which farmers accept and grow new cultivars will depend in large part on the decisions early in a breeding program on specific traits and objectives and the diligence with which those traits are pursued in a selection and testing program. To a lesser degree, the choice of a breeding method appropriate to the crop and traits of interest will influence success. Most breeders believe that the genetic material going into a program, the care with which it is handled, and the subsequent testing procedures have a greater impact on eventual productivity and adaptation of new cultivars than the use of any specific crossing and selection strategy. Testing and introduction of new cultivars into current cropping systems provide an immediate evaluation of their value to the producer, putting the genetic package into perspective as part of a total strategy to improve productivity, profitability, or other objectives of farmers and their systems.

At some point, it may become apparent in a breeding program that the genetic potential of the cultivars in use is greatly limiting productivity. This was the case in the Restrepo region of Colombia in the early 1970s where on-farm trials revealed that introduced black bean varieties were 30 to 50% higher yielding than the traditional large-seeded red or mottled varieties. Further analysis of markets showed that farmers were rational in their growing of traditional varieties because of a large price advantage for the mottled bean varieties in local markets. Trials and demonstrations designed to show benefits from new varieties and fertilizer and insecticide applications had little impact on the adoption of any of these components. Rather, the farmers learned from the trials the advantages of higher plant densities and planting in rows to aid in cultivation—two limitations not identified by the researchers or extension specialists at the outset. Later analysis showed that the latter factors were far more limiting to the production of beans by the local small farmers, given their inability to purchase fertilizer or insecticides and the market preference for one type of bean. This experience underscores the importance of a strategy that clearly identifies limiting factors and designs research to deal with the entire system including the economic realities of the farmers and the environment in which they operate. A well-designed and carefully implemented crop improvement program would do little to improve productivity if there were other major constraints in the system.

Information on the relative importance of genetic improvement versus other factors can be gleaned from the testing program, especially if a wide range of locations is used for evaluation. If these locations are on farms or are accessible to farmers, there is opportunity to get farmers' feedback on the desirability of the introduced cultivars compared to current ones and their perceptions on what constraints are most limiting. Field tours, farmer meetings, and local cooperatives or other organizations can all be helpful in evaluating the appropriateness of new technologies, including new cultivars. Breeders and other research or extension specialists must address limitations to productivity as perceived by farmers, or demonstrate through participatory activities on the farm that the farmers' perceptions are not consistent with biological or economic reality at that site. Otherwise the

73

research will have a low likelihood of payoff, at least in the short term. It is more likely that there will be a number of factors that limit productivity or profit and that these are interrelated. For this reason, research in sustainable agriculture has most promise of success if situations in the field are approached by an interdisciplinary team that is capable of identifying and evaluating a range of factors and dealing with several dimensions of the farming system: biological, economic, social, and environmental.

On-farm testing of component and alternative systems is one approach to involving clients and learning how farmers view their systems and make decisions about inputs and technologies. This does not mean merely locating "experiment station type" trials on farmers' fields as a convenient way to add environments to a testing program. The practical application of on-farm research includes identifying the farmer on the research and extension team as a full member of the team, from identification of problems to design of alternatives to evaluation of results. In plant breeding, this most often takes the form of a comparison of introduced cultivars with local or currently accepted cultivars. The large-scale, full-sized equipment approaches using long test strips, replications, and randomization of entries provide highly acceptable data for the evaluation of cultivars. In one set of trials over several years in Clay County, Nebraska, a comparison of maize hybrids on three or four farms each year with a single replication per farm consistently gave coefficients of variation of 3 to 4%, often lower than similar trials on the experiment stations (Rzewnicki et al., 1988). This research/ demonstration strategy can supplement the replicated trials conducted by breeders under more controlled conditions on station.

In summary, the development and eventual adoption of new cultivars, both hybrids and varieties, are complex challenges faced by plant breeders in both developed and developing countries. Choice of a crossing method appears to be less critical than careful choice of objectives, parent materials, and testing under the conditions that are expected to prevail when the new cultivars are released from the program. Plant breeding has made large contributions to increased crop productivity, but genetic improvement does not operate in isolation. Improvement of crop cultivars for sustainable pro-

74

duction systems requires careful evaluation of the constraints in current systems and the projected challenges in future systems. The process requires consideration of the efficiency and cost-effectiveness of crop breeding versus other strategies of improving or maintaining productivity. Most important, the improvement of crops for these systems must involve an evaluation of the interaction between crop cultivars and other technologies in the system and the many alternatives that could be pursued. These factors should be included in the choice of objectives and methods for a crop breeding program.

Acknowledgments

The ideas and commentary on the outline for this chapter by Dr. Blaine Johnson (Univ. of Nebraska), Dr. Steve Baenziger (Univ. of Nebraska), and Dr. Tom Barker (Pioneer Hi-Bred International) are sincerely appreciated. Careful review of the final manuscript by Dr. Brett Callaway contributed greatly to the chapter.

References

Alexander, H.M., and P.J. Bramel-Cox. 1991. Sustainability of genetic resistance. p. 11–27. *In* D.A. Sleper, T.C. Barker, and P.J. Bramel-Cox (ed.) Plant breeding and sustainable agriculture: Considerations for objectives and methods. Crop Sci. Soc. Am. Spec. Pub. 18. Amer. Soc. Agron., Madison, WI.

Allen, F.L., R.E. Comstock, and D.C. Rasmusson. 1978. Optimal environments for yield testing. Crop Sci. 18:747–751.

Atlin, G.N., and K.J. Frey. 1990. Selecting oat lines for yield in low-productivity environments. Crop Sci. 30:556–561.

Baker, R.J. 1986. Selection indices in plant breeding. CRC Press, Boca Raton, FL.

Bell, A.A. 1982. Plant pest interaction with environmental stress and breeding for pest resistance: Plant diseases. p. 335–364. *In* M.N. Christiansen and C.F. Lewis (ed.) Breeding plants for less favorable environments. Wiley, New York, NY.

Blum, A. 1988. Plant breeding for stress environments. CRC Press, Boca Raton, FL.

Bradley, J.P., K.H. Knittle, and A.F. Troyer. 1988. Statistical methods in seed corn product selection. J. Prod. Agric. 1:34–38.

Bramel-Cox, P.J., T. Barker, F. Zavala-Garcia, and J.D. Eastin. 1991. Selection and testing environments for improved performance under reduced-input conditions. p.29–56. *In* D.A. Sleper, T.C. Barker, and P.J. Bramel-Cox (ed.) Plant breeding and sustainable agriculture: Considerations for objectives and methods. Crop Sci. Soc. Am. Spec. Pub. 18. Amer. Soc. Agron., Madison, WI.

Christiansen, M.N., and C.F. Lewis (ed.). 1982. Breeding plants for less favorable environments. Wiley, New York, NY.

Clark, R.B., J.W. Maranville, and H.J. Gorz. 1978. Phosphorus efficiency of sorghum grown with limited phosphorus. p.93–99. *In* A.R. Ferguson, R.L. Bieleski, and I.B. Ferguson (ed.) Plant nutrition 1978. Proc. 8th Intl. Colloq. Plan Anal. Fert. Prob., Aukland, New Zealand.

Coffman, W.R., and M.E. Smith. 1991. Roles of public, industry, and international research center breeding programs in developing germplasm for sustainable agriculture. p.1–9. *In* D.A. Sleper, T.C. Barker, and P.J. Bramel-Cox (ed.) Plant breeding and sustainable agriculture: Considerations for objectives and methods. Crop Sci. Soc. Am. Spec. Pub. 18. Amer. Soc. Agron., Madison, WI.

Devine, T.E. 1982. Genetic fitting of crops to problem soils. p.143–173. *In* M.N. Christiansen and C.F. Lewis (ed.) Breeding plants for less favorable environments. Wiley, New York, NY.

Dofing, S.M., and C.A. Francis. 1990. Efficiency of one-replicate yield testing. J. Prod. Agric. 3:399–402.

Edwards, C.A., R. Lal, P. Madden, R.H. Miller, and G. House (ed.). 1990. Sustainable agricultural systems. Soil & Water Conserv. Soc., Ankeny, IA.

Fehr, W.R. 1987a. Principles of cultivar development. Vol.1, Theory and technique. Macmillan, New York, NY.

Fehr, W.R. 1987b. Principles of cultivar development. Vol.2, Crop species. Macmillan, New York, NY.

Francis, C.A. (ed.). 1986. Multiple cropping systems. Macmillan, New York, NY.

Francis, C.A. 1990. Breeding hybrids and varieties for sustainable systems. p.24–54. *In* C.A. Francis, C.B. Flora, and L.D. King (ed.) Sustainable agriculture in temperate zones. Wiley, New York, NY.

Francis, C.A., C.B. Flora, and L.D. King (ed.). 1990. Sustainable agriculture in temperate zones. Wiley, New York, NY.

Francis, C.A., R.R. Harwood, and J.F. Parr. 1986. The potential for regenerative agriculture in the developing world. Amer. J. Altern. Agric. 1:63–74.

Jackson, W. 1980. New roots for agriculture. New ed. Univ. Nebraska Press, Lincoln, NE.

Jenkins, J.N. 1982. Plant pest interactions with environmental stress and breeding for pest resistance: Insects. p.365–374. *In* M.N. Christiansen and C.F. Lewis (ed.) Breeding plants for less favorable environments. Wiley, New York, NY.

Jensen, N.F. 1980. Crop breeding as a design science. p.21–29. *In* D.R. Wood (ed.) Crop breeding. Amer. Soc. Agron., Madison, WI.

Jensen, N.F. 1988. Plant breeding methodology. Wiley, New York, NY.

Kalloo. 1988a. Vegetable breeding. Vol.1. CRC Press, Boca Raton, FL.

Kalloo. 1988b. Vegetable breeding. Vol.2. CRC Press, Boca Raton, FL.

Kalloo. 1988c. Vegetable breeding. Vol.3. CRC Press, Boca Raton, FL.

Liebman, M., and R.R. Janke. 1990. Sustainable weed management practices. p.111–143. *In* C.A. Francis, C.B. Flora, and L.D. King (ed.) Sustainable agriculture in temperate zones. Wiley, New York, NY.

Rosielle, A.A., and J. Hamblin. 1981. Theoretical aspects of selection for yield in stress and non-stress environments. Crop Sci. 21:943–946.

Rzewnicki, P.E., R. Thompson, G.W. Lesoing, R.W. Elmore, C.A. Francis, A.M. Parkhurst, and R.S. Moomaw. 1988. On-farm experiment designs and implications for locating research sites. Amer. J. Altern. Agric. 3:168–173.

Sasser, J.N. 1982. Plant pest interactions with environmental stress and breeding for pest resistance: Nematodes. p.375–390. *In* M.N. Christiansen and C.F. Lewis (ed.) Breeding plants for less favorable environments. Wiley, New York, NY.

Sleper, D.A., T.C. Barker, and P.J. Bramel-Cox (ed.). 1991. Plant breeding and sustainable agriculture: Considerations for objectives and methods. Special Pub. 18. Crop Sci. Soc. Am., Madison, WI.

Smith, M.A., and R.W. Zobel. 1991. Plant genetic interactions in alternative cropping systems: Consideration for breeding methods. p.57–81. *In* D.A. Sleper, T.C. Barker, and P.J. Bramel-Cox (ed.) Plant breeding and sustainable agriculture: Considerations for objectives and methods. Crop Sci. Soc. Am. Spec. Pub. 18. Amer. Soc. Agron., Madison, WI.

Soule, J., and J. Piper. 1992. Farming in nature's image. Island Press, Washington, DC.

Zobel, R.W. 1975. The genetics of root development. p.261–275. *In* J.G. Torrey and D.F. Clarkson (ed.) The development and function of roots. Academic Press, London, England.

Zweifel, T.R., J.W. Maranville, W.M. Ross, and R.B. Clark. 1987. Nitrogen fertility and irrigation influence on grain sorghum nitrogen efficiency. Agron. J. 79:419–422.

6

Breeding for Resistance to
Insects and Plant Pathogens

Michael P. Hoffmann, H. David Thurston, and Margaret E. Smith

The protection of food and fiber crops from insect pests and plant pathogens in current agricultural systems relies to a large extent on the use and continued availability of synthetic pesticides. Future reliance on these chemicals, however, is doubtful. Health, safety, and environmental issues have been identified, and resistance to pesticides by insects and plant pathogens is of major concern. In addition, the regulation of pesticide use has become more stringent, resulting in fewer pesticides available and fewer new materials marketed each year.

Alternatives to conventional pesticides for control of pests are available. Crop rotation, sanitation, pest-free seed, site selection, water management, biological control, biorational pesticides, and pest-resistant crop plants are all examples of tactics that have some pest-control benefit. These tactics are components of a holistic and ecologically based strategy, generally referred to as integrated pest management. Pesticides are used in integrated pest management, but only when all other efforts to maintain pests at acceptable levels have failed. Preference is given to those tactics that are the most benign and environmentally sensitive, yet still provide adequate control.

This chapter will focus on the breeding of crop plants for resistance to insect pests and plant diseases and how this can contribute to the sustainability of modern agricultural systems. It will also emphasize the value of crops selected by traditional farmers and the current and potential contributions these genotypes make to the world's food supply.

Background and Current Status

The development of crop plants resistant to disease and insect pests is by no means new. Traditional farmers have been selecting for pest resistance for thousands of years. Many of the major crops on which humans depend for food were developed long before modern agricultural science began. Buddenhagen (1981) reminds us of the value of these crops: "Millions of acres of many crops are varieties, or landraces, or clones selected by ancient men and women in prehistory, or at least before agricultural science was developed. This is so for the *Dioscorea* yams, for most of the rice and cowpeas in West Africa, and for most of the maize and beans in Latin America. It is still true for several million acres of rice in Asia, and much of the potato crop of the Andes in South America. . . . Sorghum and millet in tropical Africa are largely old landraces, as is much of the forage grass acreage of the world."

Landraces are usually genetically diverse and in balance with the environment and endemic pathogens. They are dependable and stable in that although not necessarily high yielding, they yield some harvest under all but the worst conditions. Cultivar selection and maintenance by traditional farmers still continue in some indigenous societies. Jennings (1976), who developed the high-yielding rice cultivar IR8 that began the Green Revolution in rice, stated: "The breeding methods devised by Neolithic Man remained standard until the 20th century, although in recent decades they were applied more systematically and with more sophistication. The technique is called pure line selection."

Scientific breeding of plants for disease resistance probably did not begin until after the disastrous potato late blight epidemic in Ireland. Potatoes had been introduced to Europe from South America in the 16th century by the Spanish and gradually spread throughout Europe. Potatoes were especially well adapted to Ireland, and after their introduction the population of Ireland rose from about 3 million to about 8 million. When the late blight fungus was introduced into Ireland it almost destroyed the potato crop during 1845 and 1846. An estimated one million Irish died in the subsequent famine and another million and a half emigrated. The late blight epiphytotic is frequently used as an example of the danger of dependence on a monoculture,

80

the risk of genetic uniformity, and the importance of crop diversification. Similar disasters waiting to happen exist today among food crops in some areas of the world. Diversification is essential for long-term stability and sustainability of cultivated food crops.

One of the earliest and most dramatic demonstrations of the benefit of plant resistance to insects occurred in the mid-1800s when grape phylloxera was introduced into Europe from North America. Grape phylloxera attacked and devastated the susceptible European grape resulting in a near collapse of the wine industry. It was known, however, that native American grapes were highly resistant to this aphid pest. By grafting European grapes onto American grape rootstocks, phylloxera was controlled and the industry saved in the remarkably short time span of ten years.

Some effort has been made to breed pest resistance into almost all crops and today host plant resistance holds an important position in most crop protection programs. Recent statistics indicate that over 650 cultivars used in the United States and elsewhere have been reported to possess resistance to plant pathogens (Shaner, 1991), and over 100 crop cultivars grown in the United States and over 200 grown in other major crop production areas of the world are resistant to insect pests (Smith, 1989). The use of resistant cultivars is a principal method of insect and disease control for grains such as sorghum, maize, and wheat.

Advantages and Disadvantages of Host Plant Resistance

As a tactic for improving the sustainability of agricultural systems, host plant resistance offers several important advantages. Host plant resistance is specific to a pest or complex of pests and rarely has a direct detrimental effect on beneficial organisms. Resistant cultivars are generally easily adopted because little modification of existing farming practices is needed and their use is generally compatible with other tactics of pest control. Resistant cultivars are economical because the cost of resistant seed is generally little or no higher than the cost of susceptible seed. Where crops are of low value and pesticide applications are not economical, the use of resistant cultivars may be the only solution to a pest problem. The use of resistant cultivars is environmentally sensitive because it eliminates or at least reduces the need for pesticides. The benefits of using resistant crops are frequently cumulative, that is, the suppression of the first generation of an insect pest results in fewer offspring for subsequent generations. The breeding process not only identifies resistance but also identifies susceptible material that should be avoided or used only in areas where pest pressure is not severe.

Tingey and Steffens (1991) summarized the benefits derived from the use of insect resistant cultivars in the United States. For example, resistance to the European corn borer in maize and the Hessian fly in wheat saves $388 million annually. When the developmental costs of pest-resistant crops are compared with the developmental costs of pesticides, pest-resistant crops yield a much higher return for each dollar invested (Bottrell, 1980).

Although the use of host plant resistance has many advantages in sustainable systems, disadvantages exist. A considerable amount of time is required to develop resistant cultivars. Although the exception rather than the rule, resistance to a single species of pest may be accomplished in three to five years. Resistance to multiple species can take much longer. Breeding is expensive. For example, in 1978, Plaisted estimated that it usually took 200,000 potato seedlings to find one worthy of release and one million seedlings to find a new cultivar that will eventually be used on a significant percentage of U.S. potato acreage (Thurston, 1978). The cost per seedling in 1978 was $1.00. In addition, resistance is not necessarily permanent. There are several examples of biotypes (resistance-breaking populations) of pests that are now able to damage plants that were previously resistant.

Priorities and Perspectives

The interests of the entomologist or plant pathologist may be of low priority relative to agronomic considerations such as yield; quality; flavor; color; texture; tolerance to abiotic stress; and transportation, harvest, and storage characteristics. Ideally, interdisciplinary teams should be formed to identify and prioritize the characters needed in new crop cultivars.

The perspective of the entomologist or plant pathologist may vary because of subtle but important distinctions in their definitions of resistance. This can result in different approaches to solving the pest problem and can inhibit interdisciplinary breeding efforts. For the purposes of this chapter we elect to use the definition of resistance proposed by Schafer (1974), which appears to apply equally well to plant pathogens and pest insects and states: "Resistance may be considered as any character that causes a plant to have less disease, insect attack, or overall loss than another." Resistance is a relative term and is measured relative to a known susceptible plant. In addition, plant breeders, entomologists, and plant pathologists frequently use different terminology when discussing plant resistance. Harris and Frederiksen (1984) provide an inventory of such terminology.

Entomologists most commonly use the terminology proposed by Painter (1951) to describe the mechanisms of plant resistance to insects. According to Painter, resistance consists of three genetically controlled and interrelated components: nonpreference, antibiosis, and tolerance. Nonpreference is the plant's ability to lessen the likelihood of recognition and use by insects. Antibiosis describes the adverse effects of a plant on insect survival, development, and/or reproduction, and tolerance is the ability of a plant to withstand insect infestations and to support populations that would severely damage susceptible plants. The effects of these three components are interrelated, but they are independent genetic characters.

Painter (1951) also proposed the term *pseudoresistance* to describe resistance that is the result of transitory characters in otherwise susceptible hosts. This type of host plant resistance includes host evasion (pest and host asynchronous), escape (lack of infestation because of chance), and induced resistance. Induced resistance is the temporary increase in resistance resulting from some condition of the plant or environment such as a change in soil temperature, soil fertility, moisture, or earlier damage.

The terminology used by plant pathologists to describe resistance to disease is far more variable and extensive than that used by entomologists. For example, there are 16 different terms used for general resistance to late blight of potato (Thurston, 1971). Some terms widely used by plant pathologists deserve discussion.

Escape as used by plant pathologists is roughly equivalent to Painter's definition of escape and/or host evasion. For example, some cultivars of wheat are early maturing and consequently avoid late season rusts. *Tolerance* also applies equally well whether in reference to insect pests or plant pathogens; however, the potential for developing damaging reservoirs of insects or inoculum may be greater with the latter. Agrios (1988) defines tolerance as follows: "The ability of a plant to sustain the effects of a disease without dying or suffering serious injury or crop loss." *Induced resistance* is also similar to that defined by Painter, except to plant pathologists, it is more related to an earlier infection by a pathogen than to environmental factors. *Immunity* describes the condition of a plant that is completely free of infection by a given disease even under conditions most favorable for the pathogen. *Hypersensitivity* is similar to immunity, except that when the pathogen invades the host, very rapid death of plant cells at the site of infection isolates the pathogen, preventing further spread into the host. The invading pathogen may die because of the death of the host cells or because of the action of host plant toxins released at the site of infection.

Insect/Pathogen Host Plant Interactions

The probability of developing plant cultivars resistant to pests is improved when the physiological and/or behavioral interactions between the pest and potential host plant are understood. The investigation of these interactions may involve several disciplines including biochemistry, behavior, and neurophysiology.

The dispersal abilities, size, physical association with host, and generation time of insects and plant pathogens are distinctly different and can for the most part explain the differences in their interactions with host plants. Harris and Frederiksen (1984) list differences between arthropods (insects and mites) and pathogens relative to host location and use. Pathogen dispersal is relatively passive, whereas insects actively search for and remain on host plants. Because insects are larger and more metabolically active than pathogens, they interact with a greater diversity of chemical and physical stimuli. Understanding and manipulating these stimuli, especially those associated with potential host plants, is critical in the development of resistance to insect pests.

The chemical and physical plant components that intervene at each step in host plant location and selection by insects have been widely investigated and their manipulation targeted for plant breeding programs. Insects follow a fairly stereotypic series of steps in discriminating and using host plants. This process includes: (1) host habitat finding, (2) host location, (3) host recognition, (4) host acceptance, and (5) host suitability. Modification (through breeding) of appropriate stimuli at any step reduces the likelihood that the insect will ultimately locate and infest the plant.

Chemical stimuli involved in host finding and recognition can be classified into two major categories. *Allomones* are compounds that result in a negative response by the insect to the plant and include repellents, feeding and oviposition deterrents, and toxicants. For example, glucose esters in glandular trichomes of wild tomato deter settling by the potato aphid (Goffreda et al., 1989). Conversely, *kairomones* are compounds released by the plant that result in a positive response by insects and include attractants, stimulants, arrestants, and excitants. Corn rootworms and cucumber beetles, for example are highly attracted to wild Cucurbitaceae. Chambliss and Jones (1966) and others demonstrated that the compounds responsible were cucurbitacins. Through breeding programs, the amount of cucurbitacin in commercial cucurbit cultivars has been greatly reduced, resulting in reduced attack by cucumber beetles.

Physical components of the host plant that may influence the selection and successful establishment of the pest include: succulence, toughness, epidermal hairs, and surface waxes. One of the more common physical components implicated in plant resistance to insects is the presence of trichomes (epidermal hairs). Glandular or secretory trichomes confer resistance by physical entrapment or through the action of toxins or other allelochemicals present in the secretions. Nonglandular trichomes act by capture or impalement. Physical resistance mechanisms are especially valuable because insects are unlikely to overcome these mechanisms, since major changes in pest anatomy and behavior are required. This type of plant resistance is also somewhat easier to research since phenotypes possessing the desirable traits are relatively easy to identify.

Compared with insects, the interactions of plant pathogens with their host plants are limited spatially, and because the dispersal phase of most plant pathogens is passive, complex behavioral interactions with the host plant do not occur. For a disease to appear, the infective pathogen must arrive at a susceptible host and the environment must be favorable. A change in the pathogen, host, or environment will affect the expression of the disease. Upon arrival at a potentially susceptible host and given a favorable environment, both physical and chemical defenses of the plant must be overcome by the pathogen.

Physical defenses include leaf and fruit waxes that form a hydrophobic surface, preventing the retention of water needed for germination. A thick cuticle or epidermal layer may also prevent successful infection, especially by fungal pathogens that rely on mechanical pressure to penetrate the host. Thick internal barriers act to prevent the spread of bacterial, fungal, and nematode pathogens. Insects that vector pathogens may be repelled by trichomes. Pathogens may also invade through natural openings in plant tissue such as stomata or lenticels. Stomata on some cultivars of wheat open late in the day. Germ tubes of fungal spores that germinated during the night desiccate before the stomata open, preventing invasion by the pathogen. Lenticels on some apple cultivars are very small and prevent entry of the apple spot bacterium. Seed coats are generally effective physical barriers against pathogen invasion, and root caps and mucilage protect outer root walls from soil-borne pathogens.

Following attack, plants may form additional physical barriers such as cork layers around the site of infection, preventing spread of the pathogen, or toxic compounds secreted by the pathogen to other tissue. Infected leaf or fruit tissue may simply fall off because of the development of abscission layers in response to the disease. Other responses of plants to attack by pathogens such as the deposition of gums, swelling of cell walls, and enveloping of fungal hyphae function to isolate or impede the pathogen.

Biochemical (preexisting or induced) defenses by plants to pathogen attack include compounds that inhibit pathogens either in the immediate vicinity of the plant or once in the plant tissue. Plants frequently respond to attack by synthesis or increased production of phenolic compounds (common phenolics or phytoalexins), which are toxic to or inhibit the growth of plant pathogens. These compounds not only defend the plant against disease but may also defend against insect attack. Plant pathologists have been aware of this concept for the past several decades; interest by entomologists is more recent. Smith (1989) lists several crop plants in which induced resistance to insects has been demonstrated. Despite the evidence that plants may respond similarly to induced resistance to either insects or pathogens, most breeding efforts are specific for either insects or pathogens.

Genetic Nature of Resistance

Knowledge of the genetic basis for resistance is invaluable to breeding programs because it dictates how to use resistance sources most efficiently in crossing and progeny selection, and facilitates the development of isogenic lines for studying the mechanisms of resistance.

The genetic nature of resistance may be monogenic (single gene, race specific, qualitative, vertical), oligogenic (a few genes), or polygenic (multigenic, general, horizontal, partial, quantitative). For certain pests, monogenic resistance often breaks down a few years after a new cultivar has been introduced. Among plant diseases, this occurs most commonly with fungal pathogens that produce tremendous numbers of individual propagules that are easily dispersed in the air, allowing for the rapid spread of favorable mutants (those that overcome resistance). For example, at the height of a potato late blight epidemic, a 100-m^2 plot of potatoes can produce as many as 300 million spores per day resulting in a 2500-fold increase in the fungal population. There are relatively few cases of resistance breakdown for viruses, bacteria, nematodes, or soil-borne pathogens.

Resistance that is independent of biotypes or races is generally polygenic and is considered to be more stable than monogenic resistance. According to Plaisted (1966), polygenic resistance has the following attributes: (1) many genes contribute small and individually indistinguishable effects upon

87

the phenotype, (2) interactions between the different genes are quite likely, and (3) the expression of the genotype is generally modified greatly by the environment. Although polygenic resistance is complicated and difficult to breed for, it tends to be long-lasting and thus provides "sustainable resistance."

The breakdown of plant resistance to insect pests is not uncommon. Biotypes of insect species are now able to successfully damage and survive on plants that were previously resistant. These populations constitute the single most serious limitation to the permanence of resistance and they occur in at least 14 insect species, 9 of which are aphids (Smith, 1989).

Genetic Vulnerability and Resources

In the United States, most major crops are dominated by a few cultivars, which means they are relatively genetically uniform. This poses a serious danger if exotic or resistance-breaking pests appear for which no genetic resistance exists. Uniformity is not necessarily dangerous unless it affects pest resistance or abiotic stress tolerance. This can be discerned, however, only through the appearance of the resistance-breaking pest or the unusual environmental stress, by which time the damage due to excessive uniformity has already occurred. Cowling (1978) listed 18 studies warning against the dangers of genetic uniformity in our major crops but noted that these studies and their recommendations have not been adequately heeded. The genetic vulnerability of crops in the United States was illustrated by the southern corn leaf blight epidemic of 1970. A severe epidemic on maize galvanized plant pathologists and breeders into a furious flurry of finger-pointing at genetic vulnerability. The Texas male sterile cytoplasm found in almost all U.S. maize before this epidemic was not thought to be susceptible to any pathogen. Since then, plant pathologists and plant breeders have made more serious attempts to incorporate diversity into modern breeding programs. Nevertheless, the crops in farmers' fields today in developed countries like the United States are far from diverse, and many cultivars are highly uniform and closely related to other cultivars on the market.

Erosion of genetic resistance to pests, the result of a small number of new crop cultivars replacing a large number of landraces, is a major concern today in the international agricultural community. This rapid change to newer high-yielding cultivars of wheat, rice, and maize threatens the survival of cultivars selected for centuries by indigenous populations. Thus, many of the invaluable genetic resources embodied in landraces that were developed in a long evolutionary process may eventually disappear. The greatest numbers of genes for resistance to pests are usually found in landraces where the host and pest have coexisted for long periods of time. Termed the Green Revolution, the rapid spread and use of high-yielding cultivars (HYVs) of wheat and rice have been spectacular. For example, over 75% of the rice area in the Philippines is planted to HYVs developed since 1965 (Herdt and Capule, 1983). The area planted to the HYVs of wheat and rice in Asia and the Near East increased from 49,000 ha in the 1965/66 crop year, to almost 55 million ha in 1976/77 (Dalrymple, 1978). In 1980, almost 40% of the rice produced in South and Southeast Asia consisted of HYVs, and they were estimated to contribute 4.5 billion dollars annually to the value of rice produced in Asia (Herdt and Capule, 1983). The rapid adoption of the HYVs of wheat and rice is one of the major success stories in the history of agriculture (Coffman and Bates, this volume, Chapter 2); however, there is considerable concern that the widespread use of a few cultivars with common genes may increase the risk of crop losses from new races of existing pathogens or currently obscure plant diseases or insect pests.

Considering the importance and magnitude of the task, efforts to conserve traditional crop germplasm have been sadly inadequate to date. Although international centers funded by CGIAR (Consultative Group in International Agricultural Research) are making efforts to conserve germplasm of a few major crops, their efforts are only a small part of a worldwide problem of maintaining diversity in many crops (Brewbaker, this volume, Chapter 8). In the long run the CGIAR centers may make one of their major contributions by collecting and safeguarding diversity in the germplasm of the world's major food crops. Survival of the genetic diversity existing in the many traditional landraces and their wild relatives, most frequently found in developing countries, is seriously threatened by human activities. Loss of these irreplaceable reservoirs of genetic material would constitute a tragedy for mankind. There are hopeful signs that the value of such germplasm is beginning to be appreciated, but financial support for its preservation is still meager.

In an interesting discussion of diversity and genetic vulnerability, Brown (1983) suggested that genetic diversity *per se* does not necessarily provide insurance for a species against genetic vulnerability. For example, in the 1950s, southern rust of maize was introduced into regions of Africa where it spread rapidly on the open-pollinated maize varieties of the region. The local landraces of maize were highly variable but also highly susceptible to southern rust and severe epidemics resulted. Brown made the important point that although genetic diversity may be highly desirable, it is of little value unless it includes genetic resistance to specific organisms. Brown concluded his article by stating, "Plant germplasm is among the most essential of the world's natural resources. Its conservation merits far greater attention than it is now receiving."

Factors Influencing the Expression of Resistance

The expression of resistance to pests can be greatly influenced by abiotic (light, temperature, relative humidity, and soil nutrients) and biotic factors (insect age or sex or plant height or density) (Tingey and Singh, 1980; Shaner, 1991). For example, wild tomatoes grown under long daylength have increased resistance to the tobacco hornworm (Kennedy et al., 1981), and light intensity has been shown to affect late blight lesion size on potato (Victoria and Thurston, 1974). Manipulation of shade is seldom considered a plant disease management practice, but for some crops, especially tropical crops, management of shade or light is important in determining yields and disease severity. Shade was almost universally used in the culture of coffee until the latter half of this century, and some tea was also grown under shade. One example of the effect of shade on disease severity is the reduced severity of brown leaf spot of coffee on shaded coffee compared with coffee grown in full sun (Thurston, 1991).

Temperature and soil nutrients may also influence the expression of resistance. Walker (1965) gives examples of resistance to pink root of onion. Generally, increases in nitrogen result in increased vegetative biomass, which provides more tissue of higher quality to plant-feeding insects. For example, powdery mildew in wheat is more severe at high levels of nitrogen (Shaner, 1973). Increases in potassium and phosphorous usually have the opposite effect as those reported for nitrogen.

Biotic factors such as the sex and age of insects are known to affect feeding rates and/or preferences. Female insects generally eat more than males to meet the requirements for egg production and food consumption, although this may vary with age. The choice of plant and feeding site may also be influenced by the sex of the insect; for example, females and males may respond differently to plant odors. It may be important to classify insects by sex and age to increase the accuracy of the resistance evaluation.

Temporal and spatial activity patterns of insects must also be considered when evaluating for resistance. The abundance or location of an insect on a host plant may vary with season, time of day, or stage of plant growth. Generally, when evaluating for resistance, pest densities near an economic level are adequate. For rapid screenings of large numbers of plants, greater densities are frequently used. The conditions under which insects are maintained before use in resistance evaluations often influence their responses.

Planting densities of crops have important effects on insect and disease incidence and severity. Fery and Cuthbert (1972) showed that a 15-fold increase in plant density of cowpea resulted in a 300% increase in damage by cowpea curculio. Dense plant stands generally increase disease incidence but in some cases (especially with virus diseases) may reduce disease. Crop density can be altered by manipulating the planting and by pruning, thinning, trellising, staking, harvesting plants or plant parts, and managing fertilizer and water applications (Palti, 1981). Host density is further manipulated by intercropping; one reason for the common use of intercropping by traditional farmers may be its important role in disease management.

The above described abiotic and biotic factors can have significant effects on the expression of resistance to insect pests and plant pathogens and explain the importance of using standardized procedures wherever possible and of closely monitoring environmental parameters during the evaluation process (Goodman and Souza et al., this volume, Chapters 3 and 10, respectively). They also explain the importance of artificial infestation or inoculation of plants with pests to ensure accurate identification of resistance.

Interactions between Plant Resistance and
Biological and Cultural Control

Plant resistance, biological control, and various cultural controls are the cornerstones of pest management in sustainable farming systems. By combining these control tactics, genetic response by the pest population to any one selection pressure is reduced. Unfortunately, relatively little effort has gone into investigating how these pest management tactics interact. For reviews on how plant resistance to insects influences the biological control of insect pests, see Bergman and Tingey (1979), Herzog and Funderburk (1985), and Boethel and Eikenbary (1986).

Studies have shown that resistant plant genotypes may influence rates of pest insect parasitism and predation, parasite and predator development, survivorship, and emergence; and infection of pest insects by fungal or bacterial pathogens. Insect pests on resistant host plants often experience increased mortality, decreased fecundity, retarded growth rates, or extended developmental rates, resulting in a reduction in nutritional quality available to their parasites and predators. Natural enemies of insect pests can also be exposed through their prey to plant toxins. Campbell and Duffey (1979) showed that alkaloids in tomatoes that have antibiotic properties against the tomato fruitworm were also highly toxic to the larval stage of its endoparasite, *Hyposoter exigua.*

Breeding plants to enhance the effect of structural barriers such as hooks or glandular trichomes on tomatoes, or tighter husks on sweet corn, has the overall effect of prolonging the time during which the pest remains exposed and susceptible to attack by natural enemies. Breeding for tolerance to pest attack has the greatest potential for compatibility with biological control because pests that are tolerated at moderate to high levels support populations of natural enemies. Potentially this would provide insect pathogens with sufficient host material to develop epizootics.

Given the apparent advantages of using resistant plants in combination with biological control, it is unfortunate that more effort is not directed at investigating the interactions between natural enemies, prey, and host plants. Even less effort has been directed at investigating the interactions between resistant cultivars and cultural controls. One can readily envision examples of the use of resistant (or even susceptible) cultivars in conjunction with cultural control. For example, the effectiveness of trap crops that act to concentrate pest populations for localized control could be enhanced by breeding for cultivars that are attractive to insects and diseases.

Developing Resistant Cultivars

In developing pest resistant cultivars, breeders have at their disposal the same array of established methods that would be applied to any breeding objective. The most appropriate method will be determined by the nature of the pest problem, availability and nature of resistance sources (elite cultivars, landraces, wild relatives, or unrelated species), genetic basis of resistance (mono-, oligo-, or polygenic), screening methods available, and financial and labor resources.

The nature of the pest problem must first be assessed to determine how pest resistance objectives mesh with the other objectives essential to a successful breeding program. Adaptation, productivity (biological and/or economic), and quality are inherent objectives in any breeding program. Whether one or several pest resistances should receive greater priority than these other objectives depends on the severity of the pest problem. Decisions on priorities among breeding objectives are also influenced by the availability and nature of resistance sources. Resistance that has been identified in locally adapted, elite genetic materials will be incorporated into a commercially competitive cultivar much more quickly and at less cost than resistance from a landrace or wild relative, which is likely to be poorly adapted locally and/or to carry numerous agronomically deleterious traits (Goodman, this volume, Chapter 3).

The agronomic value and local adaptation of resistance sources will influence breeding methodology. Elite, well-adapted sources could be used efficiently through pedigree, bulk, or single-seed-descent methods. Unadapted, agronomically poor-quality sources would be better subjected to backcross or recurrent selection procedures. Whether resistance in a particular source is mono-, oligo-, or polygenic will affect priorities among different breeding objectives and choice of methods as well. The more genes involved in controlling resistance, the more complex, costly, and long term the breeding effort required to incorporate that resistance into a commercially competitive cultivar. For resistance that is simply inherited, methods that are effective for traits of relatively high heritability can be used (such as mass selection, pedigree, and backcross). Polygenic resistance must be dealt with via methods (such as recurrent selection) that are effective for traits of low heritability.

The ease, cost, and precision of evaluations for pest resistance determine the breeding methods used. For certain pathogens, screening can be done in a controlled environment such as a growth chamber or greenhouse. For other pathogens and most insects, controlled-environment screening is not correlated with field responses or is impractical. In these cases, breeders must evaluate resistance based on family performance in several replications, to minimize the influence of field variation. The more reliable the pest infestation technique (resulting in fewer escapes and lower plant-to-plant variation), the smaller number of plants will be required for an accurate evaluation of each family.

For any breeding objective, the choice of breeding method is limited by the resources available. This limitation is often particularly severe for pest resistance breeding, because laboratory rearing of pest organisms is expensive and labor intensive, the influence of field and pest population variation on expression of resistance is often large, and the process of rating thousands of plants using what are often complex subjective scales is lengthy and labor intensive.

Recent advances in molecular genetics offer tools that, in some cases, may make the process of breeding for pest resistance more efficient (Mc-

94

Couch et al., this volume, Chapter 9). Molecular markers, such as restriction fragment length polymorphisms (RFLPs), are usually not influenced by environment thus having heritabilities of 1.0. If markers linked to the genes conditioning resistance in a particular source can be identified, they can be used to screen segregating progenies rather than expensive field evaluations. Molecular marker technology is presently limited by the identification of linkages with genes conditioning important resistances, by the difficulty of dealing with traits conditioned by more than one or a few genes, and by its expense. On-going research is addressing these limitations. Molecular markers, however, cannot identify new sources of resistance. As pest populations evolve, any given resistance mechanism may become less effective, or ineffective. Only field screening with pests representative of those in the natural population can address this problem.

Transfer of resistance genes from species that are sexually incompatible with the host plant can increase the genetic variation available to breeders. Once transferred to the desired species resistance genes can be manipulated by the breeder using the same array of breeding methods discussed above. Field evaluation of resistance always will be required to assess the effectiveness of resistance genes and the agronomic value of the cultivar carrying them.

Powerful new technologies such as molecular marker-based selection or gene transfer will need to be used cautiously to avoid intensive selection pressure on pest populations resulting in resistant races or biotypes. Lessons learned from the rapid breakdown of certain types of pest resistance, and epidemics resulting from extensive genetic uniformity with respect to genes affecting host-pest interactions, must be carefully considered to avoid repeating past mistakes with these new technologies.

Role of Host Plant Resistance in Sustainable Agriculture

Host plant resistance is among the oldest methods of pest control and has proved to be a very important and preferred tactic for the management of pests. It is also one of the least expensive and safest of pest control tactics and its development and use principally involve renewable resources. Pest-resistant cultivars possessing appropriate agronomic qualities can make a major contribution to the sustainability of agriculture.

Pest-resistant cultivars may function as the primary pest control tactic or complement other tactics such as biological, cultural, or chemical control. They are used in both smaller, traditional, and large, modern agricultural systems. Their use can greatly simplify pest management, especially if they function as the primary control tactic. From a pest management standpoint, their primary ecological and economic benefit is to reduce the need for pesticide inputs. Consequently, their use results in less human and environmental hazard and less potential for the development of resistance to pesticides.

The impact pest-resistant cultivars have in an agricultural system will depend on several factors, and determining their appropriateness involves a holistic examination of the agroecosystem. The pest or pest complexes (both pathogen and insect) present and their relative importance must be known. The probability of successful control using resistant cultivars should be ascertained and current control tactics, their costs, and effectiveness considered. Alternatives that may be equal to, or better than, host plant resistance must also be considered. If breeding for a resistant cultivar is justified, one must consider the type and level of resistance to deploy. The success and stability of the resistant cultivar will depend on the mode and level of resistance deployed. Cultivars that are resistant to multiple pests are preferred.

With pesticide options becoming more limited, a greater proportion of pest control will depend on tactics such as host plant resistance. Improvements in defining breeding objectives based on multidisciplinary inputs will improve the efficiency of future efforts to develop pest resistant cultivars.

Acknowledgments

We thank Dr. Ward Tingey, Department of Entomology, Cornell University, for a critical review of this chapter. We also thank Westview Press, Boulder, CO, for granting permission to use material from *Sustainable Practices for Plant Disease Managment in Traditional Farming Systems,* by H.D. Thurston, 1991.

References

Agrios, G.N. 1988. Plant Pathology. Academic Press, New York, NY.

Bergman, J.M., and W.M. Tingey. 1979. Aspects of interaction between plant genotypes and biological control. Bull. Entomol. Soc. Amer. 25:275–279.

Boethel, D.J., and R.D. Eikenbary. 1986. Interactions of plant resistance and parasitoids and predators of insects. Ellis Horwood, Chichester, England.

Bottrell, D.R. 1980. Integrated pest management. Council of Environmental Quality. U.S. Government Printing Office, Washington, DC.

Brown, W.L. 1983. Genetic diversity and genetic vulnerability: An appraisal. Econ. Bot. 37:4–12.

Buddenhagen, I.W. 1981. Conceptual and practical considerations when breeding for tolerance or resistance. p.221–234. *In* R.C. Staples and G.H. Toenniessen (ed.) Plant disease control: resistance and susceptibility. Wiley, New York, NY.

Campbell, B.C., and S.S. Duffey. 1979. Tomatine and parasitic wasps: Potential incompatibility of plant antibiosis with biological control. Science. 205:700–702.

Chambliss, O.L., and C.M. Jones. 1966. Cucurbitacins: Specific insect attractants in Cucurbitaceae. Science 153:1392–1393.

Cowling, E.B. 1978. Agricultural and forest practices that favor epidemics. p.361–381. *In* J.G. Horsfall and E.B. Cowling (ed.) Plant disease: An advanced treatise. Vol.2, How disease develops in populations. Academic Press, New York, NY.

Dalrymple, D.G. 1978. Development and spread of high-yielding varieties of wheat and rice in the less developed nations. Foreign Agric. Econ. Report. No. 95. Econ. Res. Serv., USDA, Washington, DC.

Fery, R.L., and F.P. Cuthbert. 1972. Association of plant density, cowpea curculio damage, and *Choanephora* pod rot in southern peas. J. Am. Soc. Hort. Sci. 97: 800–802.

Goffreda, J.C., M.A. Mutschler, D.A. Ave, W.M. Tingey, and J.C. Steffens. 1989. Aphid deterrence by glucose esters in glandular trichome exudate of the wild tomato, *Lycopersicon pennellii*. J. Chem. Ecol. 15: 2135–2147.

Harris, M.K., and R.A. Frederiksen. 1984. Concepts and methods regarding host plant resistance to arthropods and pathogens. Ann. Rev. Phytopathol. 22:247–272.

Herdt, R.W., and C. Capule. 1983. Adoption, spread, and production impact of modern varieties in Asia. International Rice Research Institute. Los Banos, Philippines.

Herzog, D.C., and J. E. Funderburk. 1985. Plant resistance and cultural practice interactions with biological control. p.67–88. *In* M. A. Hoy and D.C. Herzog (ed.) Biological control in agricultural IPM systems. Academic Press, New York, NY.

Jennings, P. 1976. The amplification of agricultural production. Sci. Am. 235:180–194.

Kennedy, G.G., R.T. Yamamoto, M.B. Dimock, W.G. Williams, and J. Bordner. 1981. Effect of daylength and light intensity on 2-tridecanone levels and resistance in *Lycopersicon hirsutum* f. *glabratum* to *Manduca sexta*. J. Chem. Ecol. 7:707–716.

Painter, R.H. 1951. Insect resistance in crop plants. Univ. Press of Kansas. Lawrence, KS.

Palti, J. 1981. Cultural practices and infectious crop diseases. Springer-Verlag, Berlin, West Germany.

Plaisted, R.L. 1966. Methods of breeding potatoes for factors affecting processing quality. p.103–123. *In* Proc. of the plant science symposium. Campbell Institute for Agric. Res. Camden, N.J.

Schafer, J.F. 1974. Host plant resistance to plant pathogens and insects: History, current status, and future outlook. p.238–247. *In* F. G. Maxwell and F. A. Harris (ed.) Proc. of the summer institute on biological control of plant insects and diseases. Miss. Agric. Expt. Sta., Mississippi State University, MS.

Shaner, G. 1973. Evaluation of slow-mildewing resistance of Knox wheat in the field. Phytopathology 63:867–872.

Shaner, G. 1991. Genetic resistance for control of plant disease. p.495–

540. *In* D. Pimentel (ed.) Handbook of pest management in agriculture. 2nd ed. CRC Press, Boca Raton, FL.

Smith, C.M. 1989. Plant resistance to insects: A fundamental approach. Wiley, New York, NY.

Thurston, H.D. 1971. Relationship of general resistance: Late blight of potato. Phytopathology 61:620–626.

Thurston, H.D. 1978. Potentialities for pest management in potatoes. p. 117–136. *In* E.H. Smith and D. Pimentel (ed.) Pest control strategies. Academic Press, New York, NY.

Thurston, H.D. 1991. Sustainable practices for plant disease management in traditional farming systems. Westview Press, Boulder, CO.

Tingey, W.M., and S.R. Singh. 1980. Environmental factors influencing the magnitude and expression of resistance. p.89–113. *In* F.G. Maxwell and P.R. Jennings (ed.) Breeding plants resistant to insects. Wiley, New York, NY.

Tingey, W.M., and J.C. Steffens. 1991. The environmental control of insects using plant resistance. p.131–155. *In* D. Pimentel (ed.) Handbook of pest management in agriculture. 2nd ed. CRC Press, Boca Raton, FL.

Victoria, J.I., and H.D. Thurston. 1974. Light intensity effects on lesion size caused by *Phytophthora infestans* on potato leaves. Phytopathology 64:753–754.

Walker, J.C. 1965. Use of environmental factors in screening for disease resistance. Annu. Rev. Phytopathol. 3:197–209.

7

Crop Tolerance to Weeds

M. Brett Callaway and Frank Forcella

As scientists begin to understand why one species is more competitive than another, we can then breed for such characteristics in crop plants. —Baldwin and Santlemann (1980)

Plant breeders have successfully increased crop tolerance to many pests and physiological stresses. Very little work, however, has examined the potential for increasing levels of tolerance to weeds despite more than $6.2 billion spent annually to control weeds on crop and pasture land, including $3.1 billion for herbicides alone (Shaw, 1982; Pimentel et al., 1991). Herbicides constitute 69% of the total U.S. pesticide use (Pimentel et al., 1991). In cropping situations where herbicides are either unavailable, prohibitively expensive, or unpopular, methods for maintaining yield while reducing the number of weedings, thereby reducing time and labor costs, would be extremely beneficial. Table 7.1 provides examples of the considerable labor requirements for weeding selected crops.

Crop cultivars with increased tolerance to weeds may reduce the time, labor, and expense devoted to weeding, as well as crop yield losses and weed propagule production. Our purposes in this chapter are to establish the potential for increasing crop tolerance to weeds and suggest some approaches to selecting genotypes with increased tolerance to weeds. A case study is provided to demonstrate one approach to selection that has been success-

Table 7.1. Examples of labor requirements for weeding selected crops in various cropping systems.

Country	% of Total Labor	Crop	System
Mali	25	paddy rice	irrigated
Ghana	28	paddy rice	irrigated
Peru	25	paddy rice	irrigated
India	25	jute	irrigated
India	41	cotton	irrigated
Sudan	31	groundnut	irrigated
Kenya	26	sugarcane	perennial crops
Sri Lanka	27	pineapple	perennial crops
Kenya	30	coffee	perennial crops
Brazil	39	coffee	perennial crops
Thailand	55	rubber	perennial crops
Malaysia	46	coconut	perennial crops
Kenya	42	cashew	perennial crops
Kenya	59	mango	perennial crops
Ghana	31	maize	fallow
Kenya	26	maize	fallow
Colombia	55	manioc	fallow
Senegal	59	pearl millet	permanent upland
Senegal	32	groundnut	permanent upland
Thailand	29	maize	permanent upland
Brazil	44	soybean	permanent upland
Kenya	40	potato	permanent upland

Source: Adapted from Ruthenburg, 1980.

fully applied. Finally, we briefly consider the possible responses of weed populations to crops with greater tolerance to weeds. We do not discuss varietal differences in allelopathic weed suppression as this has been reviewed elsewhere (Rice, 1984; Putnam and Tang, 1986). Similarly, breeding for resistance to parasitic weeds has also been reviewed recently (Ramaiah, 1987; Obilana, 1987) and is not discussed. Curiously, the more universal phenomenon of nonallelopathic varietal tolerance to weeds has largely been ignored by plant breeders.

Evidence for Genotypic Differences in
Tolerance to Negative Influences from Neighboring Plants

What is the relation between competitive ability in plants and resistance to weeds?
Sakai (1961)

This section is included not as an attempt at a thorough review of the topic but rather to point to more exhaustive reviews of the subject and to highlight evidence from various disciplines. Although there is voluminous literature on genotypic differences in tolerance to negative influences from neighboring plants, curiously, its application to weed management has been lacking. Instead, alternative weed management practices have emphasized differences in response to abiotic factors. Abiotic factors are often more easily controlled, simplifying their study. This is a long-standing research bias, as noted by Darwin (1859): "We have reason to believe that species in a state of nature are closely limited in their ranges by the competition of other organic beings, and that there is a deeply-seated error of considering the physical conditions of a country as the most important [for its inhabitants]; whereas it cannot be disputed that the nature of the other species with which each has to compete, is at least as important, and generally a far more important element of success." The following paragraphs provide evidence from the literature supporting the contention of genotypic differences in tolerance to neighboring plants. Once this is established, we can move to the next step—the exploitation of these differences for improving agricultural production methods.

CROP-WEED SYSTEMS: GENOTYPIC DIFFERENCES IN
CROP TOLERANCE TO WEEDS

Callaway (1990) cited a number of studies of crop varietal tolerance to weeds. Examples of varietal differences in tolerance to weeds were found for 17 of 18 crops. Some of the larger studies are highlighted below.

In a study of 25 cultivars of rice, Kawano et al. (1974) found intraspecific competition and spacing response to be highly correlated with competitive ability with weeds. Characters of importance were high vegetative vigor, large leaf area, and high rate of nitrogen absorption in the early stages of growth. More specifically, plant weight at an early stage of growth and plant height were correlated most significantly with competitive ability, whereas tillering was not found to be important.

Guneyli et al. (1969) evaluated 48 hybrids and 41 parental lines of sorghum for their tolerances to weeds. 'SD 441', 'RS 609', 'OK 613', 'TX 660', and 'KS 602' were most competitive with weeds and had the least yield loss as a result of weed infestation. Competitive advantage was determined to be largely attributable to rapid germination, early emergence, and root and shoot growth during early development.

Rose et al. (1984) initially screened 280 accessions of soybeans for weed tolerance. From these, twenty were selected for further study. In general, competitive ability increased with later maturity. This was suggested to be the result of the better seed quality of later maturing accessions, which increased germination percentage and vigor. Further examination of this data (Forcella, unpublished) revealed a more complex relationship between maturity, competitive ability, and seed quality.

Challaih et al. (1983) evaluated 85 winter wheat cultivars using number of weeds as the criterion for competitive ability. At similar weed densities, cultivars were identified as having significantly greater yields than the widely grown 'Centurk 78'. It was noted that more weeds emerged in cultivars that intercepted less light. Characters of importance were tiller number, plant height, plant height index, wheat dry weight, and seedling growth rate.

Other crops listed by Callaway (1990) included alfalfa, barley, common bean, maize, cotton, cowpea, mungbean, guineagrass, pigeonpea, potato, ryegrass, squash, and sugarcane. A number of characters were identified or suggested as improving varietal tolerance to weeds. Depending on the crop species, associated weed species, and cropping practices, a given character may have a positive, neutral, or negative effect on varietal tolerance to weeds. As a result, the choice of selection environment and definition of target environment become extremely important (see below).

CROP-CROP SYSTEMS

Tolerance of Crops to Associated Crop Species. The literature on this subject has become quite large and will not be reviewed here. Much of this work has been discussed in reviews by Francis (1985), Smith and Francis (1986), Gomez and Gomez (1983), and several symposium proceedings (International Crops Research Institute for the Semi-Arid Tropics, 1981; Keswani and Ndunguru, 1982; Monyo et al., 1976; Papendick et al., 1976; and International Rice Research Institute, 1977). Several traits were identified by Francis (1985) as being particularly important in developing cultivars for multiple cropping systems: (1) maturity, (2) photoperiod sensitivity, (3) temperature sensitivity, (4) morphology, (5) root system, (6) seedling growth rate, and (7) density response.

Tolerance of Crops to Conspecifics. Genotypic differences in tolerance to negative intraspecific influences from neighboring crop plants have been recognized and used by plant breeders for many years (Harlan and Martini, 1938; Suneson and Wiebe, 1942; Eberhart et al., 1964; Jensen and Federer, 1965). Bulk-population breeding is a breeding method that makes use of these genotypic differences (Allard, 1960, and Jensen, 1988, provide more detailed discussions). Genetically diverse populations of self-pollinating crops are planted together, "in bulk," and natural selection is allowed to take place for several generations. Those genotypes that are favored by the selection environment associated with a particular bulk become predominant and, assuming a positive correlation between survival ability and agronomic worth, are selected by the plant breeder.

NATURAL SYSTEMS AND PASTURES

Kelley and Clay (1987), working with two perennial grasses, poverty oat-grass and sweet vernalgrass, found that the competitive performance of a given genotype often depended on the genotypic identity of the competing species. They concluded that the interactions were complex and had an important genetic component. Turkington and Harper (1979) sampled clones of white clover (TRFRE) that had grown in association with one of four grasses—colonial bentgrass (AGSTE), crested dogtailgrass, common vel-vetgrass, or ryegrass (LOLPE)—for many years. The clones were identified as to which grass they were associated with, such as $TRFRE_{(AGSTE)}$. They were transplanted into a monoculture sward of each grass such that every combination of $TRFRE_{(original\ grass)}$ by monoculture grass sward was represented. On average, each TRFRE type performed best when grown in association with the grass species with which it originated. Similar results were obtained in a subsequent study (Aarsson and Turkington, 1985) involving four TRFRE-LOLPE neighboring pairs from different localities. TRFRE genets (genetic individuals) generally performed best when grown with the LOLPE genet with which it was originally associated. Linhart (1988) demonstrated differentiation within a population of purslane speedwell (VERPE). Plants sampled from an area subject to high densities of almost pure stands of VERPE performed better (produced more seeds) when grown under conditions of intraspecific competition than under interspecific competition. Conversely, plants sampled from an area subject to interspecific competition performed better when grown under conditions of interspecific competition than under intraspecific competition.

Case Study: Soybean

Considerable research has been conducted on intraspecific variation within soybean for competitive ability with weeds (CAW[1]). Several studies (Burnside, 1972; McWhorter and Hartwig, 1972; McWhorter and Barrentine,

1. In this study, a high CAW genotype is defined as one that produces high yield in the presence of weeds and/or low weed yields. This is expressed as $CAW_i = v(Var_i/Var_{mean})^* w(Weed_{mean}/Weed_i)$, where VAR_i is the yield of crop variety i in the presence of weeds, VAR_{mean}

1975; Burnside and Moomaw, 1984; and Rose et al., 1984) have identified cultivars having high CAW. Moreover, soybean is one of the most extensively planted crops in North America, with over 90% of its acreage receiving one or more annual herbicide applications. Consequently, soybean is an ideal crop species with which to explore breeding potentials for high CAW.

The following section will provide a case study of a soybean breeding program for high CAW. The objective of this program was not to develop a successful commercial cultivar but to explore the possibility of breeding specifically for CAW, while holding maturity and yield equivalent to a standard, locally adapted cultivar. There were two components to this research program: first, determining which crop traits govern CAW; and second, breeding specifically for such traits.

DETERMINATION OF IMPORTANT TRAITS

Methods. In 1985 through 1987, four to eight soybean cultivars (Table 7.2) were planted in late May or early June in Stevens Co., MN. In each year there was an equal number of cultivars with high and low CAW, previously identified from Burnside and Moomaw (1984) and Rose et al. (1984). Cultivars represented plots, and weedy and weed-free treatments represented subplots, all arranged in a randomized complete block (RCB) design with 4 replications.

Beginning two weeks after emergence, periodic determinations were made on percentage emergence, leaf area per plant (LAP), leaf weight per plant, leaf number per plant, shoot weight per plant, stem weight per plant, plant height, branch number, petiole length, leaf area density, specific leaf weight, leaf area per branch, and leaf allocation (percentage dry weight allocated to leaves). At harvest, seed yield per plant and seed yield reduction were determined. Weed biomass was also determined at least three times (early, mid-, and late season) in each plot and is the variable upon which CAW is based.

---

is the average yield of all varieties under comparison in the presence of weeds, $WEED_i$ is the weed biomass associated with variety i, $WEED_{mean}$ is the average of all varieties under comparison, and v and w are weighting coefficients.

Table 7.2. Characteristics of eight soybean cultivars used in experiments at Morris, MN.

Cultivar	Competitive Ability (CAW)	Crop Yield	Weed Yield
Cumberland	Low	—	High
Century	High	—	Low
Evans	Low	Low	—
Gnome	High	—	Low
Manchu	Low	Low	—
Pixie	High	High	—
Sprite	High	High	—
Will	Low	—	High

Note: Values for characteristics were determined in screening trials at the University of Nebraska (Burnside and Moomaw, 1984; Rose et al., 1984). The characteristics are overall competitive ability, soybean yield in the presence of weeds, and weed yield in the presence of soybean.

Linear regressions were performed on weed biomass and soybean traits at each sampling date. Variability (r^2) in each regression attributable to a soybean trait was calculated. Finally, seasonal trends in r^2 values were examined.

Results. Significant relationships between weed biomass and soybean characteristics were rarely observed. Where they did occur, they appeared random and did not form trends, with the exception of LAP. LAP was the primary character that correlated consistently and significantly with weed biomass. Interestingly, a seasonal trend occurred in this relationship (Fig. 7.1). Although LAP is apparently related to weed biomass throughout the growing season, the intensity of the relationship changes during the season. Upon soybean emergence, r^2 of the LAP-weed relationship is about 0.5 but rises to a maximum of > 0.9 in early July, and subsequently drops to

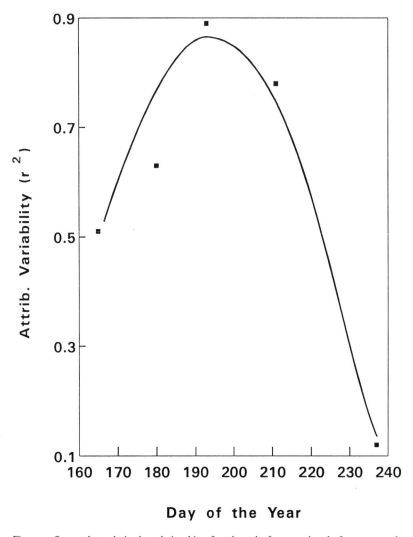

Fig. 7.1. Seasonal trends in the relationship of soybean leaf area and end-of-season weed biomass.

The r^2 values represent variability of linear regressions attributable to soybean leaf area (independent variable) when weed biomass is the dependent variable. At each date, regression calculations were based on 16 field plots of weed-stressed soybeans: four plots each of cultivars 'Century', 'Evans', 'Gnome', and 'Will'.

insignificance in late August. Although a competitive soybean requires a relatively high LAP soon after emergence, high LAP is most critical approximately 4 to 6 weeks after emergence, when crop-weed competition is most intense (see Fig. 7.1). During anthesis and thereafter, high LAP becomes progressively less important regarding crop-weed interaction, probably because weed competition has already exerted its effects on the crop.

In conclusion, high early-season LAP, that is, high leaf-area expansion rate (LAER), appears to be the trait most closely linked with CAW. Using isogenic lines of tall fescue that differed only in LAER, Forcella (1987) confirmed that the genes that code for LAER also confer CAW. The next step was to breed soybean for high LAER and determine if such a breeding effort resulted in high CAW progeny.

SELECTION FOR LEAF AREA EXPANSION RATE

F_1 and F_2 Generations. 'Evans' (soybean maturity group 0) and 'Gnome' (soybean maturity group 2) were chosen for study. These cultivars differed in CAW (see Table 7.2) and in LAER (Fig. 7.2). 'Evans' is widely grown in central Minnesota and adjacent North and South Dakota (ca. 45–47° N). 'Gnome' cannot be grown reliably in this same area because of its longer maturity; instead, it is adapted to a latitudinal band of about 42–44° N from Ohio to Nebraska.

The cultivars were sown sequentially in a greenhouse and crossed in 1986. Greenhouse-reared F_1 progenies were selfed. Approximately 1000 F_2 progenies were hand-planted in May 1987 in a nursery in Stevens Co., MN. Leaf areas of individual plants during the first six weeks of growth were calculated by measuring apical leaflet length of all trifoliate leaves on a plant and relating these measurements to leaf area per trifoliate through a previously determined allometric relationship. Leaf areas were summed for all leaves on each plant for calculation of LAP.

Leaf-areas of F_2 siblings varied considerably. The histogram in Fig. 7.3a represents a typical frequency of LAP values among F_2 progeny from a single F_1 parent. The frequently correlated trait branch number per plant is shown in Figure 7.3b. F_2 plants were selfed and seeds were selected from

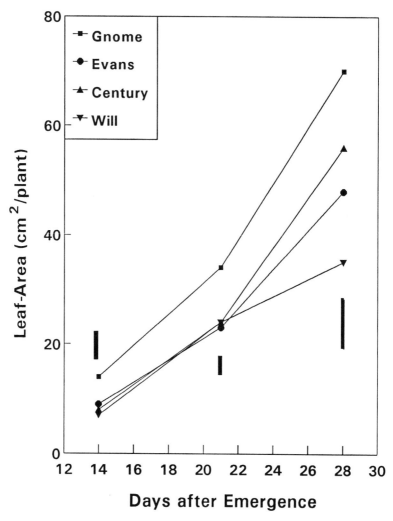

Fig. 7.2. Leaf areas of four soybean cultivars during the first four weeks of growth in a greenhouse.

Greenhouse conditions were as follows: 14 h photoperiod with light intensity of 800 mE/m^{-2}/s; day/night temperatures of 25/20 C; and soil continuously moist. Least significant difference, lsd (p<0.05) based on ANOVA, are represented by vertical bars.

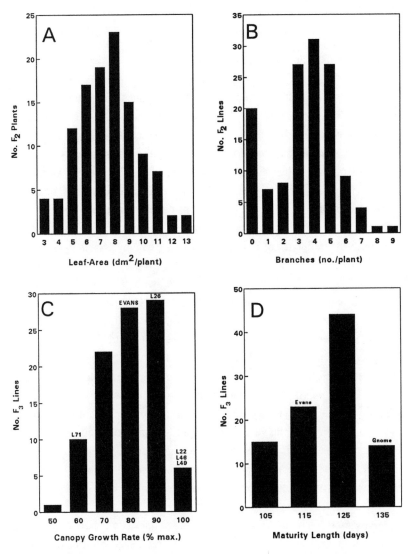

Fig. 7.3. Frequency histograms of traits in F_2 and F_3 progeny of a cross between 'Evans' and 'Gnome' soybean.

A. Variation in leaf-area per plant among F_2 individuals arising from a single F_1 parent.

B. Variation in branch number per plant among F_2 individuals arising from a single F_1 parent.

C. Variation in canopy growth rate among F_3 lines. The bars of the histogram with which 'Evans' and lines L22, L26, L46, L49, and L71 were associated are noted in the figure.

D. Variation in maturity among F_3 lines. The bars of the histograms with which 'Evans' and 'Gnome' were associated are noted in the figure.

plants with either high or low LAP values and whose seed yields were equal to or higher than that of 'Evans'.

F_3 *Generation.* In 1988, 50 seeds of each of 90 F_3 lines were sown in single-row plots. Three sets each of 'Evans' and 'Gnome' seeds were similarly planted. A commercial light meter was used to measure sunlight interception (SI) by soybean canopies five times during the first six weeks in all 96 plots. Average SI values, which are indexes of canopy growth, were adjusted to represent percentage maximum SI at each sampling date. The dates represented replications. These replications were averaged and an LSD was calculated based on the analysis of variance. The histogram in Figure 7.3c represents the frequency of F_3 lines with relative canopy growth rates (based on percentage maximum SI) ranging from 50 to 100% of the maximum rate observed. The rate for 'Evans' was about 80%, with 36 lines having greater canopy growth rates. The canopy growth rate of 'Gnome' was about 90%.

Days to maturity (Fig. 7.3d) and yield of each F_3 line were also determined. We needed to develop a soybean line that could grow under Minnesota conditions. That is, it had to have a maturity length similar to that of 'Evans'. Consequently, any genetic line whose maturity was greater than that of 'Evans' was discarded; only short maturity lines were selected. Lines whose maturities and yields were equivalent to those of 'Evans' and whose mean canopy growth rate values were one LSD (p=0.05) unit greater than 'Evans' were selected. These lines were L22, L26, L46, and L49. Additionally, L71 was selected for comparative purposes. This latter line had a very low relative canopy growth rate (Fig. 7.3c).

F_4 *Generation.* F_4 seeds from the five selected lines (L22, L26, L46, L49, and L71) as well as 'Evans' were grown in replicated 20-liter pots in a greenhouse (conditions as in Fig. 7.2) with and without competition from weeds. Weedy pots contained eight weed seedlings each of redroot pigweed and giant foxtail. Additionally, the same soybean lines were sown in weedy and nonweedy field plots (RCB design with four replications) in early June 1990. Weedy plots consisted of natural infestations of green foxtail and

Table 7.3. Comparison of traits between greenhouse-grown 'Evans' soybean and five F_4 lines.

Trait	Evans	L22	L26	L46	L49	L71
			Soybean Line			
(a) WDW[1] (g/pot)	73	121*	45*	93	70	99
(b) Y_{WF} (g/pot)	15	14	19	18	18	17
(c) Y_{WS} (g/pot)	7	5*	9**	6	7	7
(d) $Y_{\%E}$(%)	44	32*	57**	36	48	44
(e) HT (cm)	83	88	101	87	108	98
(f) SWT (g/plant)	23	21	29	27	27	26
(g) BN (no./plant)	2	2	5***	4**	4*	3
(h) PN (no./plant)	37	31	42	39	38	37
(i) SN (no./plant)	80	68	96	83	78	83

Note: Compared traits are (a) dry weights of competing weeds, WDW; (b) weed-free soybean yields, Y_{WF}; (c) weed-stressed soybean yields, Y_{WS}; (d) yield as percentage of weed-free 'Evans', $Y_{\%E}$; (e) height of weed-free soybean at harvest, HT; (f) shoot dry weight of weed-free soybean, SWT; (g) branch number of weed-free soybean, BN; (h) pod number of weed-free soybean, PN; and (i) seed number of weed-free soybean, SN. Statistical probabilities based on t-tests of differences between 'Evans' and other lines are indicated by the following superscripts:*, $p<0.10$; **, $p<0.05$; and ***, $p<0.01$.
[1]WDW of weed growing in the absence of soybean was 147 g/pot.

redroot pigweed. For both greenhouse and field experiments, t-tests were used to detect differences between 'Evans' and each of the selected F_4 lines for each trait.

Based on the greenhouse experiment, one of the selected lines, L26, had an appreciably higher CAW than that of 'Evans'. Two sources of evidence permit this conclusion. First, the weight of weeds grown with L26 was only 45 g/pot, whereas that for 'Evans' was significantly greater, 73 g/pot (Table 7.3). Dry weights of weeds competing with other lines were similar to or

Table 7.4. Seed yields per plant of field-grown 'Evans' soybean and five experimental soybean lines in both weedy and weed-free plots.

Soybean	Yield (s.d.) (g/plant)	t-statistic	p-value
Weedy Plots			
L22	8.7 (1.7)	−0.19	0.86
L26	11.6 (1.1)	1.73	0.14
L46	8.1 (2.3)	0.14	0.90
L49	7.8 (2.1)	0.28	0.79
L71	7.4 (2.3)	0.44	0.68
Evans	8.3 (3.0)		
Weed-Free Plots			
L22	10.6 (3.1)	−0.39	0.72
L26	11.2 (1.4)	−0.96	0.37
L46	8.9 (1.9)	0.66	0.53
L49	8.5 (1.4)	0.86	0.43
L71	9.3 (3.4)	0.28	0.79
Evans	9.9 (2.4)		

Note: Means, standard deviations (s.d.), t-statistics from t-tests, and probability (p) values are provided for comparison.

greater than that of 'Evans'. Second, the yield of weed-stressed L26 was 8.8 g/pot, which was significantly greater than that of 'Evans', 6.8 g/pot. Yields of no other lines were significantly greater than 'Evans'. The high yield of weed-stressed L26 was not the result of an overall increase in vigor or yield components. For example, the yields of L26 and 'Evans' under weed-free conditions did not differ significantly (Table 7.3); nor did plant height, shoot dry weight, or numbers of pods and seeds per plant differ significantly between these two genotypes. Branch number per plant was significantly greater in L26 than in 'Evans', however, and this trait may have been a contributing factor in the high CAW of L26, although it apparently was not for L46 and L49.

In the 1990 field trial, weedy plot yield of L26 soybean (11.6 g/plant) was greater than that of 'Evans' (8.3 g/plant) (Table 7.4). Differences between yields of 'Evans' and other lines were smaller, and p-values greater, than that for L26 and 'Evans'. In weed-free plots, the probability of a difference between seed yields of 'Evans' and any other line being the result of chance was very high.

Although the agronomic performance of L26 is unknown, the greenhouse and field results indicate that purposeful breeding for high CAW is possible in crops such as soybean. The trait responsible for high CAW appears to be high LAER. High branch number per plant in soybean may also contribute directly or indirectly through LAER to high CAW. Almost certainly many additional characteristics may elicit or modify a genotype's CAW, but these characteristics have yet to be studied specifically. Consequently, considerable opportunity appears to exist for increasing the efficiency of current crop cultivars and simultaneously increasing the number of nonchemical options for weed management in agriculture.

This case study illustrates one approach to screening for tolerance to weeds that has been successful. A number of other approaches are discussed below. In initiating a selection program for identifying crop genotypes with improved tolerance to weeds, a plant breeder must carefully consider how to fit this program into the objectives and time frame of the overall breeding effort. It is hoped that the strategies that follow will provide the breeder with a framework from which to proceed.

Strategies for Selecting Crop Genotypes with Tolerance to Weeds
APPROACHES TO SCREENING FOR TOLERANCE TO WEEDS

How competitive ability or tolerance to weeds is defined is of critical importance to the success of screening and/or breeding programs. A number of definitions of competitive ability and tolerance have been used in the literature (see Callaway, 1990), usually with no justification for their choice. Frequently the way in which competitive ability is measured is never described. The definition required may vary according to the time-frame of the particular project. Early in the development of cultivars, selection for tolerance to weeds may emphasize traits such as rapid emergence, photoperiod sensitivity, or maturity. At this stage "tolerance" may be defined relative to these traits rather than to yield *per se*. This reflects the common practice among plant breeders of selecting for traits conditioned by one or a few genes early in the selection process. As the selection process progresses, the number of lines to be evaluated is fewer and more emphasis is placed on quantitative traits, particularly yield. Evaluations at this stage frequently are replicated. At this stage, *tolerance* may now require a definition of "those lines which are the highest yielders under conditions of weed infestation," since these materials are closer to being placed on the market.

Another way of approaching the problem of developing weed tolerant crops is in terms of genotype by environment interaction (GxWE, where WE is the weed environment). This provides an established theoretical framework from which to begin (Souza et al., this volume, Chapter 10). In a discussion of natural systems, Kelley (1985) states, "Not only do different species constitute different environments but different *genotypes* [italicized for emphasis] of another species constitute different environments for a particular species." This same concept has been stated by Francis (1981) and Gomez and Gomez (1983) in developing cultivars for use in multiple cropping systems. Using this framework, is a significant interaction desirable? It seems that this question alone is too simple to yield a useful answer. For a thorough discussion of considerations in using and interpreting GxWE, the reader is referred to Souza et al. (this volume, Chapter 10) and references therein.

GxWE is the interaction of many genotypic traits with the weed environment. In the simplest case, an interaction may be the result of a single trait interacting with a single environmental variable, such as biomass partitioning and photoperiod. The expression of this single trait may have a dominating influence on the expression of the trait of ultimate interest, such as yield. The plant breeder must give careful consideration to just which trait is actually being selected. Considering grain yield as the dependent variable, we may attribute the response of this variable to several components, such as partitioning (that is, Harvest Index), biomass yield, growth rate, and tolerance to weeds. These components may be confounded with each other as illustrated by a hypothetical example. A set of genotypes exhibiting a significant GxWE are of the following composition: Those with (1) greater sensitivity to weed stress, leading to an earlier partitioning to reproductive organs, a shorter vegetative growth period (lower biomass yield), and lower total grain yield; (2) more tolerance of weed stress, leading to continued vegetative growth (greater biomass yield), later partitioning to reproductive organs, and higher total grain yield; (3) an "average" response to weed stress, but greater partitioning and smaller overall biomass (shorter maturity); (4) an average response to weed stress, but smaller partitioning and greater overall biomass (later maturity); and (5) "average" partitioning but differences in rates of biomass accumulation (growth rates), leading to stress avoidance by higher growth-rate genotypes (shading weeds, better root development, and so on).

The above are distinct (though frequently correlated) mechanisms leading to GxWE interactions. Selection for stability of response (lack of a significant GxWE interaction) may be desirable or undesirable, depending on the mechanism responsible for the response and the target environment for the selected cultivar. This example illustrates how genetic improvement for yield stability (stress tolerance) may suffer from a poor understanding of the mechanisms of significant GxWE interactions. By the same argument, a nonsignificant GxWE may be misleading because of the canceling effects of component traits conferring "nonstable" performance over environments. Eisemann et al. (1990) discuss this problem in greater detail. With careful attention to interpretation, however, GxWE can provide valuable information for the selection of more weed-tolerant crop genotypes.

Assuming a suitable weed environment (see discussion below), several methods for screening crop cultivars have been suggested for intercrops that may be adapted to crop-weed systems. First, it must be determined whether or not selection under weedy conditions is necessary, that is, do genotypes perform differently relative to each other under weedy and weed-free conditions? Francis (1981), Gomez and Gomez (1983), and Smith and Francis (1986) have suggested several methods for making this determination. Correlations of genotype ranks between no-weed and weedy environments may be used. This method has the disadvantage that even low correlation coefficients may be highly significant, thus situations may arise in which a large percentage of total variation is unaccounted for by the correlation. A second method involves the use of an analysis of variance. In this method the sum of squares is partitioned into a GxWE component. A significant interaction term in which more than two weed environments are involved may be further partitioned into genotype by *no-weed vs. weedy* and genotype by *among weedy* terms. A significant genotype by *no-weed vs. weedy* term indicates that the genotypes perform differently relative to each other in the two environments. In this event, the genotypic performance may be regarded as not one trait but two (performance under weedy and weed-free conditions) that are genetically correlated (Falconer, 1952). A third technique involves the use of a Finlay and Wilkinson (1963) type of regression (see Souza et al., this volume, Chapter 10, for a detailed discussion). A regression is computed for the performance of each genotype in a particular weed environment vs. the mean performance of all genotypes at each weed environment. The resulting regression coefficient is used to compare different genotypes over weed environments (Gomez and Gomez, 1983). Significant differences indicate that separate selection environments are needed for genotypes targeted for no-weed and weedy environments.

CHOICE OF SELECTION ENVIRONMENT

A desirable selection environment would have two characteristics: (1) Wide application, that is, genotypes selected in this environment as being tolerant should exhibit this characteristic in the target environment(s). Commercially, genotypes that perform well over many environmental conditions (many target environments) are preferred. (2) Ease of establishment; that is, complex or expensive methods required to establish the selection environment may make selection for weed tolerance economically unattractive.

Several studies suggest methods by which these characteristics may be achieved at least partially. Turkington (1979) demonstrated that genotypes of alfalfa originally growing in association with timothy were more productive when planted with other species than were other alfalfa genotypes. This result points to the possibility of using *tester species* for the evaluation of crop tolerances to weeds. A tester species could be planted in association with crop genotypes to evaluate their tolerances to weeds in much the same manner as tester lines are used by maize breeders to evaluate the heterotic potential of inbred lines. The concept of a tester species implies that competitive abilities among weed species differ only in magnitude and that interactions of weed species with crop genotypes are insignificant. Although this is not likely to be a universal occurrence, it may be common enough for many important weed-crop systems.

If suitable tester species are identified for even a few important systems, selection for crop tolerance to weeds may be greatly simplified. For example, if lambsquarters and redroot pigweed are found to be similar in their competitive abilities, they may be considered as one species for the purpose of selection. In addition, species differing only in magnitudes of competitive abilities may be considered in "competitive equivalents" (Goldberg and Werner, 1983; Pantone and Baker, 1991). That is, one redroot pigweed plant of (for example) biomass b may equal two of lambsquarters, each with a biomass equal to b^{-2}. This approach will aid in determining the types of target environments to which weed-tolerant crop genotypes may be adapted because weed communities consisting of many species may be treated, for selection purposes, as consisting of a single "competitive load."

A second approach to the problem of choosing the selection environment is to apply the methods frequently used by plant breeders to group testing environments, specifically to weed environments. A number of methods have been proposed (Abou-El-Fittouh et al., 1969; Brown et al., 1983; Crossa et al., 1991; Horner and Frey, 1957; Zobel et al., 1988). The combined use of an AMMI (additive main effects and multiplicative interaction) model and a biplot summarization permits the simultaneous examination of main effects for genotypes, main effects for environments, and their interction (Crossa et al., 1990a; Gauch, 1985 and 1988; Zobel, 1990; Zobel et al., 1988). Weed environments that exhibit consistent main effects (selection pressure relatively consistent among years) and high interaction (effective separation of genotypic tolerance to weeds) would be preferred and may be identified using this method. (See Herben and Krahulec [1990] and Mortensen and Coble [1989] for examples of complications caused by year-to-year fluctuations.)

Various combinations of these approaches may be used. For example, tester species may be identified through the use of statistical grouping of weed (species) environments using techniques such as the AMMI/biplot method. Similarly, the "competitive equivalencies" of weeds may be determined using some form of multivariate statistical grouping as described by Crossa et al. (1990b). In addition, several mathematical models (Pantone and Baker, 1991; Spitters and Van Den Bergh, 1982; Vandermeer, 1989; Watkinson, 1985) may be used to identify tester species and determine competitive equivalents.

METHODS OF GENOTYPE IDENTIFICATION

Several methods have been proposed for identifying genotypes for use in intercrop conditions. Willey and Rao (1981) suggested the use of a regression similar to that of Finlay and Wilkinson (1963) to predict performance of genotypes of one crop when intercropped with different genotypes of another crop. For our purposes the associated weed community may be considered the "other crop." This regression is familiar to plant breeders and is carried out as described above (see "Approaches to Screening for Tolerance to Weeds"). Assuming that the responses (or their transformed values) can be described adequately by a linear relationship, the slope of a genotype's regression indicates its tolerance (or conversely, its sensitivity) to weeds. A slope of zero indicates that the genotype is totally unaffected by weeds. For the purpose of this discussion all slopes will be assumed to be negative. As the slope increases, the relative sensitivity of the genotype to the weed environment becomes greater. A slope greater than 1.0 indicates that the genotype is relatively more sensitive to weed environments with high mean values (biomasses, densities, and so on) than those with low

mean values. Since the objective is to identify genotypes with higher tolerance (less sensitivity) to weeds, slopes greater than 1.0 are undesirable.
Sensitivity is, of course, not the only criterion of consequence. In the extreme case of a genotype with no yield, yield would be totally unaffected by
weed environment, but the genotype could hardly be considered useful.
Therefore, the height of the regression line should also be considered.
Willey and Rao (1981) discuss this in the context of land equivalent ratios (LERs) for two associated crops. Following their suggestion, an "expected" value is calculated for a standard point on the horizontal axis. Using
0.5 as this standard point,[2] an expected value would be obtained as the
corresponding Y-axis value for the genotype of interest (as determined from
its regression equation). For a genotype with an expected value of 0.68, the
expected yield advantage would be 18% (0.68–0.5=0.18) (Fig. 7.4).

Hamblin et al. (1976) propose experimental designs that permit two
species to be selected simultaneously for yield and ecological combining
ability. The purpose of their designs is to study the segregating populations
of N varietal crosses of parental species i with n varietal crosses of parental
species j. Nn plots are required in each replicate so that each of the Nn cross
by species combinations are represented. A disadvantage of this design is
immediately apparent in that for as few as 10 crosses of each species, 100
plots are necessary per replication (for balanced complete block designs). If
comparisons are to be made with monoculture performance of each species
and/or parent, the number of plots required increases accordingly.

Federer (1979) suggested response models for evaluating mixtures of
genotypes. These models would be appropriate if crop genotypes were
being evaluated in the presence of mixtures of weeds. They include components of variance that are attributable to general and specific "mixing"

2. The Y-axis value of 1.0 is the average weed-free yield. A negative slope of 1.0 describes
an additive relationship between X and Y values. The X-axis value of 1.0 is that point at which
the average genotype yield reaches zero assuming an additive relationship between X and Y.
Therefore, 0.5 is the point at which half of the crop yield would be lost to weeds for an average-
yielding genotype responding additively to weeds. This model provides a theoretical framework from which genotype responses may be considered.

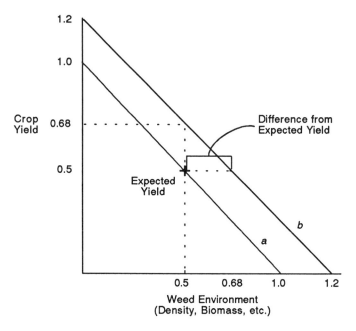

Fig. 7.4. Weed tolerance of genotype *b* relative to the "average" genotype (*a*). (See text.)

effects of crop genotypes with each other and with the weeds. Models such as this (and those discussed by Wright, 1982, and Griffing, 1976) would be useful for the simultaneous consideration of crop and weed yields and would allow estimation of relative tolerances. These models also address the question of whether or not crop genotypes that are less competitive in monoculture and in association with weeds are also poorer yielders. Depending on the breeder's objectives, genotypes could be selected for two criteria. The first criterion would be significant positive general mixing effects (gme), that is, genotypes exhibiting a positive yield response to weeds. These genotypes would be ideal if they existed, but this is unlikely to be a commonly observed response, except perhaps in very sparse weed populations. The second criterion is non-significant gme's, that is, those genotypes showing no significant differences in response to weeds. A third option, significant negative gme's, would not be desirable. These models also allow consideration of specific mixing effects (sme's), introducing the possibility of identifying crop genotypes that are particularly well suited to competition with specific weed species that may be used in situations of specific weed problems. For example, certain potato cultivars have been shown to be more tolerant of yellow nutsedge than others (Yip et al., 1974).

Weed Response

A final consideration is how weed populations respond to neighboring crop genotypes. Do weed-tolerant cultivars lead to weeds with greater competitive ability? In this context, Antonovics (1978) concluded, "Genetic changes in competitive relations have rarely been studied" and that they were "in need of investigation." Although fifteen years have passed since these remarks, the only study of which we are aware that examines the genetic response of a weed population to competition from a crop is that by Jordan (1989). Jordan made use of sib analysis to estimate quantitative genetic parameters for predicting evolutionary response of a non-weedy population of poorjoe to competition from soybean. His model predicted that the response of the non-weedy population would eventually produce a phenotype similar to the weedy population in most of the traits considered. Ramakrishnan and Gupta (1972) evaluated three ecotypes of bermudagrass for their competitive effect on the growth of wheat. They found significant differences among bermudagrass ecotypes in effects on wheat-shoot dry weight. In this study, the most competitive ecotype was also affected least by the crop. Because the ecotypes exhibited no apparent morphological differences, the authors attributed their results to eco-physiological differences among the ecotypes. We are aware of no studies examining the genetics of weed population responses to *genotypes* of a crop. From the perspective of communities of species rather than genotypes, Goodnight (1990) demonstrated how selection may act directly at the species community level, resulting in the differentiation of a community into various community phenotypes. Therefore, we return to the question, Do weed-tolerant cultivars lead to weeds with greater competitive ability? These studies establish the *potential* of weed-tolerant cultivars leading to more competitive associated weed populations, and/or more competitive weed communities. Good agricultural practices, such as crop rotation, would probably prevent or slow such selection, just as herbicide rotation inhibits selection for herbicide resistant weeds.

How could one examine the genetics of natural selection within weed populations as they adjust to the introduction of weed-tolerant crop cultivars? One way is through a combination of plant demographics and quantitative genetics. Usually the demographics (life history traits) required include an estimate of the size of the soil seed bank, censuses of each seedling cohort from emergence to death, and a measure of the seed production for each individual (Mohler and Callaway, 1992; Moore and Chapman, 1986). Although extremely labor intensive, this procedure provides a way of estimating survivorship probabilities and mortality factors for the weed species under consideration. According to Endler (1986), "This is the only method in which it is possible to obtain complete lifetime fitness estimates, which are required in order to predict changes in or equilibria of trait frequency distributions." Genetic relationships between relatives should be established in the weed population used. This design permits the use of Lande's model for measuring selection on correlated characters (Lande, 1982b; Lande and Arnold, 1983). The model is defined as $\Delta\bar{z} = GP^{-1}S$. Where $\Delta\bar{z}$ is a column vector representing the change in the mean of each phenotypic character from one generation to the next, G is the additive genetic variance-covariance matrix for the characters, P is the phenotypic variance-covariance matrix assessed before selection, $^{-1}$ signifies matrix inversion, and S is a column vector of selection differentials (Arnold and Wade, 1984b). Therefore the selection gradient, $P^{-1}S$, accounts for phenotypic correlations (positive or negative) of z_i with other characters. This analysis, "helps to reveal the target(s) of selection, and to quantify its intensity" (Lande and Arnold, 1983) and is a major improvement over the univariate selection gradients typically used by weed ecologists (Endler, 1986; but see also James and McCulloch, 1990). More detailed discussions of the theory and use of correlated characters are found in Arnold and Wade (1984a,b), Lande (1977, 1979, 1982a,b, 1988), Lande and Arnold (1983), and Endler (1986).

Tolerance as Part of an Integrated Weed Management System

From the perspective of sustainable agriculture, it may be argued that a crop must not only tolerate a given level of weed infestation, it must also reduce weed numbers, biomass, and so on. Otherwise, weed infestations will become worse in subsequent years. We feel, however, that this argument takes an overly fragmented view of crop production. The development of crops tolerant to weeds will be paralleled by other improvements in weed management techniques. Just as improvements in insect and disease resistance have not eliminated the need for alternative insect and disease management options, neither will weed-tolerant cultivars eliminate the need for alternative weed management options. As another useful tool in an integrated weed management system, however, weed-tolerant cultivars have the potential to save millions of dollars in production costs, as have insect and disease-resistant cultivars.

References

Aarssen, L., and R. Turkington. 1985. Biotic specialization between neighboring genotypes in *Lolium perenne* and *Trifolium repens* from a permanent pasture. J. Ecol. 73:605–614.

Abou-El-Fittouh, H.A., J.O. Rawlings, and P.A. Miller. 1969. Classification of environments to control genotype by environment interactions with an application to cotton. Crop Sci. 9:135–140.

Allard, R.W. 1960. Principles of plant breeding. Wiley, New York, NY.

Antonovics, J. 1978. The population genetics of mixtures. p.233–252. *In* J. Wilson (ed.) Plant relations in pastures. CSIRO, Australia.

Arnold, S.J., and M.J. Wade. 1984a. On the measurement of natural and sexual selection: Theory. Evol. 38:709–719.

Arnold, S.J., and M.J. Wade. 1984b. On the measurement of natural and sexual selection: Applications. Evol. 38:720–734.

Baldwin, F.L., and P.W. Santleman. 1980. Weed science and integrated pest management. Bioscience 30:675–678.

Brown, K.D., M.E. Sorrells, and W.R. Coffman. 1983. A method for classification and evaluation of testing environments. Crop Sci. 23:889–893.

Burnside, O.C. 1972. Tolerance of soybean cultivars to weed competition and herbicides. Weed Sci. 20:294–297.

Burnside, O.C., and R.S. Moomaw. 1984. Influence of weed control treatments on soybean cultivars in an oat-soybean rotation. Agron. J. 76: 887–890.

Callaway, M.B. 1990. Crop varietal tolerance to weeds: A compilation. Mimeo No. 1990–1. Dept. Plant Breeding & Biometry, Cornell Univ., Ithaca, NY.

Challaih, R.E. Ramsel, G.A. Wicks, O.C. Burnside, and V.A. Johnson. 1983. Evaluation of the weed competitive ability of winter wheat cultivars. North Central Weed Control Conf. Proc. 38:85–91.

Crossa, J., R.W. Zobel, and H.G. Gauch, Jr. 1990a. Additive main effects and multiplicative interaction analysis of two international maize cultivar trials. Crop Sci. 30:493–500.

Crossa, J., W.H. Pfeiffer, P.N. Fox, and S. Rajaram. 1990b. Multivariate analysis for classifying sites: Application to an international wheat yield trial. p.214–233. In M.S. Kang (ed.) Genotype-by-environment interaction and plant breeding. Louisiana State University, Baton Rouge, LA.

Crossa, J., P.N. Fox, W.H. Pfeiffer, S. Rajaram, and H.G. Gauch, Jr. 1991. AMMI adjustment for statistical analysis of an international wheat yield trial. Theor. Appl. Genet. 81:27–37.

Darwin, C. 1859. The origin of species. First Collier Books Facsimile edition, 1962. Macmillan, NY.

Eberhart, S.A., L.A. Penny, and G.F. Sprague. 1964. Intraplot competition among maize single crosses. Crop Sci. 4:467–471.

Endler, J.A. 1986. Natural Selection in the Wild. Princeton University Press, Princeton, NJ.

Eisemann, R.L., M. Cooper, and D.R. Woodruff. 1990. Beyond the analytical methodology: Better interpretation and exploitation of genotype-by-environment interaction in breeding. p.108–117. In M.S. Kang (ed.) Genotype-by-environment interaction and plant breeding. Louisiana State University, Baton Rouge, LA.

Falconer, D.S. 1952. The problem of environment and selection. Am. Nat. 86:293–298.

Federer, W.T. 1979. Statistical designs and response models for mixtures of cultivars. Agron. J. 71:701–706.

Finlay, K.W., and G.N. Wilkinson. 1963. The analysis of adaptation in a plant-breeding programme. Aust. J. Agric. Res. 14:742–754.

Forcella, F. 1987. Tolerance of weed competition associated with high leaf-area expansion rate in tall fescue. Crop Sci. 27:146–147.

Francis, C.A. 1981. Development of plant genotypes for multiple cropping systems. p.179–231. *In* K.J. Frey (ed.) Plant Breeding Symposium II, Iowa State University Press, Ames, IA.

Francis, C.A. 1985. Variety development for multiple cropping systems. CRC Crit. Rev. Plant Sci. 3:133–168.

Gauch, H.G., Jr. 1985. Integrating additive and multiplicative models for analysis of yield trials with assessment of predictive success. Mimeo 85–7. Dept. Agron., Cornell Univ., Ithaca, NY.

Gauch, H.G., Jr. 1988. Model selection and validation for yield trials with interaction. Biometrics 44:705–715.

Goldberg, D.E., and P.A. Werner. 1983. Equivalence of competitors in plant communities: A null hypothesis and a field experimental approach. Am. J. Bot. 70:1098–1104.

Gomez, A.A., and K.A. Gomez. 1983. Multiple cropping in the humid tropics of Asia. Pub. 176e. IDRC, Ottawa, Canada.

Goodnight, C.J. 1990. Experimental studies of community evolution I: The response to selection at the community level. Evol. 44:1614–1624.

Griffing, B. 1976. Selection for populations of interacting genotypes. p.413–434. In E. Pollak, O. Kempthorne, and T.B. Bailey (ed.) Proceedings of the international conference on quantitative genetics. Iowa State University Press, Ames, IA.

Guneyli, E., O.C. Burnside, and P.T. Nordquist. 1969. Influence of seedling characteristics on weed competitive ability of sorghum hybrids and inbred lines. Crop Sci. 9:713–716.

Hamblin, J., J.G. Rowell, and R. Redden. 1976. Selection for mixed cropping. Euphytica 25:97–106.

Harlan, H.V., and M.L. Martini. 1938. The effect of natural selection in a mixture of barley varieties. J. Agric. Res. 57:189–199.

Herben, T., and F. Krahulec. 1990. Competitive hierarchies, reversals of rank order, and the deWitt approach: Are they compatible? Oikos 58: 254–256.

Horner, T.W., and K.J. Frey. 1957. Methods for determining natural areas for oat varietal recommendations. Agron. J. 49:313–315.

International Crops Research Institute for the Semi-Arid Tropics. 1981. Proceedings of the international workshop on intercropping, Hyderabad, India. January 1979. International Crops Research Institute for the Semi-Arid Tropics, Hyderabad, India.

International Rice Research Institute. 1977. Proceedings, symposium on cropping systems research and development for the Asian rice farmer, Los Banos, Philippines. 21–24 September 1976. International Rice Research Institute.

James, F.C., and C.E. McCulloch. 1990. Multivariate analysis in ecology and systematics: Panacea or pandora's box? Annu. Rev. Ecol. Systematics 21:129–166.

Jensen, N.F. 1988. Plant breeding methodology. Wiley, New York, NY.

Jensen, N.F., and W.T. Federer. 1965. Competing ability in wheat. Crop Sci. 5:449–452.

Jordan, N. 1989. Predicted evolutionary response to selection for tolerance of soybean (*Glycine max*) and intraspecific competition in a nonweed population of poorjoe (*Diodia teres*). Weed Sci. 37:451–457.

Kawano, K., H. Gonzalez, and M. Lucena. 1974. Intraspecific competition, competition with weeds, and spacing response in rice. Crop Sci. 14:841–845.

Kelley, S.E. 1985. The effects of neighbors as environments: Characterisation of the competitive performance of *Danthonia spicata* genotypes. p.203–221. *In* P. Jacquard, G. Heim, and J. Antonovics (ed.) Genetic differentiation and dispersal in plants. Springer-Verlag, Berlin, West Germany.

Kelley, S.E., and K. Clay. 1987. Interspecific competitive interactions and the maintenance of genotypic variation within two perennial grasses. Evol. 41:92–103.

Keswani, C.L., and B.J. Ndunguru (ed.). 1982. Proceedings of the second symposium on intercropping in semi-arid areas. Pub. 186e. IDRC, Ottawa, Canada.

Lande, R. 1977. Statistical tests for natural selection on quantitative characters. Evol. 31: 442–444.

Lande, R. 1979. Quantitative genetic analysis of multivariate evolution, applied to brain: Body size allometry. Evol. 33:402–416.

Lande, R. 1982a. Elements of a quantitative genetic model of life history evolution. p.21–29. In H. Dingle and J.P. Hegmann (ed.) Evolution and genetics of life histories. Springer-Verlag, New York, NY.

Lande, R. 1982b. A quantitative genetic theory of life history evolution. Ecol. 63:607–613.

Lande, R. 1988. Quantitative genetics and evolutionary theory. p.71–84. In B.S. Weir, E.J. Eisen, M.M. Goodman, and G. Namkoong (ed.) Proceedings of the second international symposium on quantitative genetics. Sinauer Associates, Sunderland, MA.

Lande, R., and S.J. Arnold. 1983. The measurement of selection on correlated characters. Evol. 37:1210–1226.

Linhart, Y.B. 1988. Intrapopulation differentiation in annual plants. III. The contrasting effects of intra- and interspecific competition. Evol. 42:1047–1064.

McWhorter, C.G. and W.L. Barrentine. 1975. Cocklebur control in soybeans as affected by cultivars, seeding rates, and methods of weed control. Weed Sci. 23:386–390.

McWhorter, C.G. and E.E. Hartwig. 1972. Competition of johnsongrass and cocklebur with six soybean varieties. Weed Sci. 20:56–59.

Mohler, C.L., and M.B. Callaway. 1992. Effects of tillage and mulch on the emergence and survival of weeds in sweet corn. J. Appl. Ecol. 29:21–34.

Monyo, J.H., A.R.D. Ker, and M. Campbell (ed.). 1976. Intercropping in semi-arid areas. Pub. 076e. IDRC, Ottawa, Canada.

Moore, P.D., and S.B. Chapman (ed.). 1986. Methods in plant ecology. 2nd ed. Blackwell Scientific, Boston, MA.

Mortensen, D.A., and H.D. Coble. 1989. The influence of soil water content on common cocklebur (*Xanthium strumarium*) interference in soybeans (*Glycine max*). Weed Sci. 37:76–83.

Obilana, A.T. 1987. Breeding cowpeas for *Striga* resistance. p.243–253. *In* L.J. Musselman (ed.) Parasitic weeds in agriculture. Vol. 1, *Striga*. CRC Press, Boca Raton, FL.

Pantone, D.J., and J.B. Baker. 1991. Weed-crop competition models and response-surface analysis of red rice competition in cultivated rice: A review. Crop Sci. 31:1105–1110.

Papendick, R.I., P.A. Sanchez, and G.B. Triplett (ed.). 1976. Multiple cropping. American Society of Agronomy, Madison, WI.

Pimentel, D., L. McLaughlin, A. Zepp, B. Lakitan, T. Kraus, P. Kleinman, F. Vancini, W.J. Roach, E. Graap, W.S. Keeton, and G. Selig. 1991. Environmental and economic effects of reducing pesticide use. Bioscience 41:402–409.

Putnam, A.R., and C.-S. Tang (ed.). 1986. The science of allelopathy. Wiley, New York, NY.

Ramaiah, K.V. 1987. Breeding cereal grains for resistance to witchweed. p.227–242. *In* L.J. Musselman (ed.) Parasitic weeds in agriculture. Vol. 1, *Striga*. CRC Press, Boca Raton, FL.

Ramakrishnan, P.S., and U. Gupta. 1972. Ecotypic differences in *Cynodon dactylon* (L.) Pers. related to weed-crop interference. J. Appl. Ecol. 9:333–339.

Rice, E.L. 1984. Allelopathy. 2nd ed. Academic Press, New York, NY.

Rose, S.J., O.C. Burnside, J.E. Specht, and B.A. Swisher. 1984. Competition and allelopathy between soybeans and weeds. Agron. J. 76:523–528.

Ruthenburg, H. 1980. Farming systems in the tropics. 3rd ed. Clarendon Press, Oxford, England.

Sakai, K.-I. 1961. Competitive ability in plants: Its inheritance and some related problems. Symp. Soc. Exp. Biol. 15:245–263.

Shaw, W.C. 1982. Integrated weed management systems technology for pest management. Weed Sci. 30 (supplement):2–12.

Smith, M.E., and C.A. Francis. 1986. Breeding for multiple cropping systems. p.219–249. *In* C.A. Francis (ed.) Multiple cropping systems, Macmillan, New York, NY.

Spitters, C.J.T., and J.P. Van Den Bergh. 1982. Competition between crop and weeds: A system approach. p.137–148. *In* W. Holzner and M. Numata (ed.) Biology and Ecology of Weeds. Dr. W. Junk, Boston, MA.

Suneson, C.A., and G.A. Wiebe. 1942. Survival of barley and wheat varieties in mixtures. J. Am. Soc. Agron. 34:1052–1056.

Turkington, R. 1979. Neighbor relationships in grass-legume communities. IV. Fine-scale biotic differentiation. Can. J. Bot. 57:2711–2716.

Turkington, R., and J.L. Harper. 1979. The growth, distribution, and neighbor relationships of *Trifolium repens* in a permanent pasture. IV. Fine-scale biotic differentiation. J. Ecol. 67:245–254.

Vandermeer, J. 1989. The ecology of intercropping. Cambridge University Press, Cambridge, England.

Watkinson, A.R. 1985. Plant responses to crowding. p.275–289. *In* J. White (ed.) Studies on plant demography: A festschrift for John L. Harper. Academic Press, Orlando, FL.

Willey, R.W., and M.R. Rao. 1981. Genotype studies at ICRISAT. p.117–127. *In* ICRISAT. Proceedings of the international workshop on intercropping, ICRISAT, Hyderabad, India.

Wright, A.J. 1982. Some implications of a first-order model of inter-plant competition for the means and variances of complex mixtures. Theor. Appl. Genet. 64:91–96.

Yip, C.P., R.D. Sweet, and J.B. Sieczka. 1974. Competitive ability in potato cultivars with major weed species. Proc. Northeastern Weed Sci. Soc. 28:271–281.

Zobel, R.W. 1990. A powerful statistical model for understanding genotype-by-environment interaction. p.126–140. *In* M.S. Kang (ed.) Genotype-by-environment interaction and plant breeding. Louisiana State University, Baton Rouge, LA.

Zobel, R.W., M.J. Wright, and H.G. Gauch, Jr. 1988. Statistical analysis of a yield trial. Agron. J. 80:388–393.

8

Tree Improvement for
Agroforestry Systems

James L. Brewbaker

Trees and shrubs are the basic elements of many sustainable agricultural systems. They produce marketable foods, feeds, fibers, fuels, and fertilizers. They provide wood for construction and crafts and can offer long-term protection to crops, animals, and people as windbreaks, shade trees, and ameliorators of harsh environments. Many trees fix nitrogen to continually enrich the soil and can be hedge-managed and integrated with crops ("alley cropping") or with crops and animals ("alley farming"). It is the rare—and usually desolate—human agroecosystem that lacks trees and shrubs.

Tree improvement research for agroforestry systems is a new subject. Historically, plant breeding has focused largely on food crops for monocultural high-input systems. Breeding research with trees is very limited in comparison with that on field crops. It has focused narrowly on three types of trees:

- Temperate fruits—such as pomes and stone fruits.
- Temperate plantation trees—such as pines and poplars.
- Tropical plantation trees—such as oil palm and rubber.

Breeding of trees like these involves approaches that are quite similar to those of row crops, seeking to maximize productivity and quality (Wright, 1976).

Research in sustainable agriculture seeks instead to optimize long-term land use efficiency, and its focus is on maximizing the productivity of an entire farming system. Such research involves goals that are easily idealized to the point that few pragmatic solutions can be offered by the plant breeder who lives a finite life span! Tree breeders, however, are noted for long-range perspectives and patience. Tree improvement in the sustainable-system context is the subject of this chapter.

The Genetic Erosion of Trees and Shrubs

The loss of forest cover on our globe is a widely recognized tragedy of an overpopulated earth, tripling in population in the 20th century alone. Trees and shrubs have been removed from over half the world's forested land to support cities, farms, and pastures. The remnant forest—solid or scattered trees—will amount to about 4 billion ha by 2000 A.D. (out of 14 billion ice-free ha land surface). The loss that continues—about 11 million ha per year—is most severe in the tropics and correlates directly with the human population increase. Tropical forests will have declined by over one-half in this century alone, to fewer than 2 billion ha. In Asia, forested lands average less than 0.1 ha per person, about the size of a small city lot, or about one-tenth the area of land needed by a good subsistence farmer.

Genetic advance with trees is directly related to the available genetic variability. The loss of forested lands is thus of immediate concern to the tree breeder as a tragic loss of potentially important germplasm. Almost all tropical trees are endangered, since civilizations do not fully grasp the significance of genetic conservation until it is too late.

Prioritization of Trees and Shrubs
THE GENETIC RESOURCE

Zobel and Talbert (1984) stress five steps that must precede tree breeding programs:

- Determination of the preferred species.
- Determination of the extent of their variability.
- Development of a broad germplasm base collection.
- Identification and combination of desired traits.
- Increase of outstanding genotypes.

Prioritization is thus the first challenge of the tree/shrub breeder working in sustainable farming systems. There are many excellent fuelwood and fiber trees, a plethora of shrubs and trees that can stabilize and improve soils, a rich assortment of attractive hardwoods (particularly if viewed for veneer or chips rather than solid lumber), and many trees that appear to serve as animal fodder. First priority must be given to finding species with rich genetic resources.

A few regions of the tropics promise long-term forest genetic resources, including Australia, New Zealand, and a few tropical islands of developed countries. The great Vavilovan centers of plant evolution, however—such as Ethiopia, Guatemala, southern Russia, and southern China—are unlikely to retain much forest genetic diversity in the next century. Relatively few genera of tropical trees have been collected and preserved *ex situ* or in seed banks to undergird long-term breeding programs. Some examples are the Mesoamerican *Pinus* spp. and *Leucaena* spp., and the Australoasian *Eucalyptus* and *Acacia* spp.

Prioritizing trees for improvement is thus highly dependent on the status of forest genetic resources. Even within the more developed countries of temperate climes, (such as Europe, northeastern USA, and northern Asia) human forest replacement has impressively exhausted native genetic variability. The forests of Scotland were reduced to 3% land cover by agriculture and sheep, much as those of Ethiopia are today. Then they were extensively replaced by introduced species. The international travel of Monterey pines, neems, Scots pines, *Eucalyptus* spp., Australian pines, and many fruit trees is testament to the belief that "the pasture is always greener on the other side of the fence." A first responsibility in long-term tree improvement is thus to determine the extent and the center of genetic variability *in situ*.

RATE OF DRY MATTER ACCUMULATION

A common misconception is that woody plants are automatically slow in growth, that is, slow to accumulate dry matter. Annual plants accumulate biomass with flair, often appearing to explode as soils warm in the spring. Much of this growth is water, and it is soon finished. Tropical C_4 grasses, such as elephantgrass and sugarcane, also accumulate dry matter rapidly

Fig. 8.1. Seedless hybrid leucaenas (*L. leucocephala* x *L. esculent*) at the age of 2.5 years in Hawaii.

and provide yield targets above 30 t/ha/yr (dry weight), which few plants can match. This biomass, however, is also in a form that requires careful timing of harvest to minimize losses.

Dry matter accumulation of outstanding trees in sustainable agricultural systems ranges up to 100 t/ha (50% moisture content) annually. It is in a form of lignified biomass that is durable and that can be harvested on demand. A three-year-old tropical tree can attain 10 m of growth and yield large amounts of wood and fodder (Fig. 8.1). Wood and charcoal are thus preferred biomass energy sources and will perhaps be our globe's major resources when oil, coal, and gas reserves—themselves derived from biomass—are completely exploited.

EASE OF BREEDING

Ease of breeding has a high priority for multipurpose trees and shrubs, with preference given to species with short seed-to-seed cycles (such as one to five years). Short life cycles naturally accelerate genetic advance, which is a product of the selection index and the number of generations of selection. The selection index in turn is a product of the intensity of selection (percentage of selected plants), the heritability of the trait being selected, and the phenotypic variability in the breeding population.

Three to four generations of almost any recurrent selection method for almost any quantitative trait ensures significant genetic advance for almost any breeder. An advance in yield of 3 to 6% per cycle is highly predictable from recurrent mass selection, given a broad genetic base. Long life cycles obviously deter progress, and methods to shorten these cycles such as grafting, controlled environments, flowering hormones, and early-flowering genotypes can be sought to accelerate progress.

Durable seeds allow long-term germplasm storage, whereas the recalcitrant germination of seeds of many forest trees poses a serious barrier to genetic progress. Tropical fruit trees often have recalcitrant seeds—such as mangoes and avocados. Germplasm storage for such species is limited to expensive clonal repositories. Agroforestry genera with recalcitrant seeds that seriously limit genetic exploitation include *Inga* (a large genus of acid-tolerant tropical legumes) and the neem genus, *Azadirachta.*

Vegetative propagation plays a major role in the use of outstanding genotypes of woody plants, notable in the tree fruits. Methods include rooting of cuttings, grafting, tissue culture, and marcotting. For these methods to be practical, they must be rapid and inexpensive. A species like *Gliricidia sepium* is important in sustainable agroforestry systems because as its common name "quickstick" implies, it is easily propagated from cuttings. Once established, clones can provide highly valuable phenotypic uniformity that can enhance appearance, management, and yield. To the breeder, they also offer an extremely effective approach to the estimation of heritability and prospective genetic gains through selection.

The Effective Gene Pool

The effective gene pool for plant improvement includes all species that may be intercrossed to produce offspring. Perennial woody plant species normally have high crossability within genera, offering a large germplasm pool to the breeder. For example, *Leucaena leucocephala*, a multipurpose fast-growing tropical tree, can produce fertile hybrids with six other *Leucaena* spp. These in turn can be crossed with all of the 15 species in the genus, effectively allowing gene transfer throughout the complex (Brewbaker, 1987; National Academy of Sciences, 1984). Similar intercrossability characterizes genera such as *Acacia, Casuarina, Erythrina, Eucalyptus, Populus* and most other well-studied tree genera. The breeder wisely protects, collects, and characterizes all interfertile species of such effective gene pools.

The tree/shrub breeder thus has a special concern for the preservation and collection of genetic resources from a wide range of species related to the commercial species. It is not enough to focus on the single species in current popularity, for these have often spread around the world quite by chance. Such dissemination has often involved only a small fraction of the germplasm of the commercial species.

Tree Evolution and the Breeders' Genetic Resources

Are the genetic resources of woody perennials in any way unique? Is the tree breeder permitted thereby some unique opportunities for genetic advance? The answer to both questions is "yes." Long-lived woody plants are subjected to sustained evolutionary pressures that endow them with a remarkable stability. Compared to annual crops, they have extraordinary levels of resistance to diseases, insects, and predatory animals. They often have immense tolerance of protracted environmental stress—drought, heat, cold, wind. They often have an annoyingly high genetic diversity within populations and are only rarely found to be homozygous for anything (a few species are self-fertile, but these are almost entirely polyploids—such as *Leucaena leucocephala*. Most tree species cannot even be selfed to create homogeneity.

Trees and shrubs are long lived and able to spread the risk of failure in reproduction over many years. In many arid regions of the world, trees should theoretically not exist, if one considers only the average rainfall. In the "one year out of twenty" that the rainfall exceeds the average by a significant margin, however, young seedlings take root and the species

137

survives. Many trees of the fast-growing, secondary forest then produce long-lived seeds that remain viable in soils for decades.

Woody plants have survived only as the result of high levels of general resistance to diseases, insects, animal depredation, and stress. Ruminant animals are known to browse on over 75% of the trees in Africa. Damage is minimized, however, by their tall growth or their armor—thorns, odors, flavors, toxins. Annual plants defy disease by rapid growth and a long fallow season, whereas perennials must carry high tolerance or succumb. Viruses are notably impotent on woody plants compared to herbaceous species and are much more typical of annual herbs than of perennials—such as maize vs. its perennial ancestor, teosinte.

The breeder of annual crops in the tropics is inevitably swamped with problems of diseases and insects, whereas the tree breeder often blithely ignores them as insignificant. The honeymoon is soon over when trees are introduced into new environments in which the parasites and predators that have kept tree insects and other pests under control in their native ecosystem do not exist.

Tree Breeders' Challenges

INSECTS

It is the small, highly preyed-upon insects that probably present the tree breeder with the most serious challenges. This is especially true of sustainable agricultural systems, where low-input (such as no pesticide) practices are encouraged. Such insects include twig and bark borers, termites, leaf miners, and a host of sucking insects such as mites (actually arachnids not insects), scale insects, aphids, psyllids, and beetles. Many leguminous trees have seed yields reduced greatly by seed weevils. Mahogany and red cedar have been eliminated from major regions of tropical forests by the shoot borer. Shoot borers have dramatically reduced populations of *Eucalyptus* spp. in Asia, and psyllids have moved through the tropics to reduce leucaena fodder yields.

The co-evolution of trees with other biota often leads to a remarkable ecological balance involving the tree's genes, their insect pests, and the predators and parasites of the pest. Biological control has been the most effective approach to control many insects. Together with moderate levels of genetic resistance, the low levels of a predatory beetle (*Curinus coeruleus*) has effected control of the leucaena psyllid in much of the tropics (Fig. 8.2). A second predator and a parasitic wasp were less effective because of their own resident parasites.

Freedom from insect pests often characterizes newly introduced perennial species. When these pests find their way to the new region they are often especially severe, acting in the absence of their co-evolved predators and parasites. This movement of insect pests has been immeasurably accelerated by the advent of human air travel. New insects arrive monthly at most airports of the world.

DISEASES

Freedom from disease is never attained by plants, but most perennial woody species achieve near freedom, symptomatically at least. Many fungi become systemic in trees and shrubs without affecting the plant except in times of stress. These include a host of soil-borne pathogens, most of which have found their way around the world, such as species of *Phytophthora, Fusarium,* and *Pythium.* Leaf rusts threaten populations of pines and rosewoods; *Fusarium* wilts often decimate populations of *Albizia* and *Acacia;* stem cankers reduce the wood value of many tree species.

Genetic tolerance of pathogens is variable and often high in tree species. Notable are the variations in tolerance of highland vs. lowland species in the tropics, where these environments involve different pathogens. Highland species often appear to lack genetic tolerance to the soil pathogens of warmer soils. Similarly, species from dry regions cannot tolerate pathogens commonly found in waterlogged soils. Tree die-back is a common phenomenon. A "replant problem" exists with many trees—such as papaya—from soil buildup of root pathogens.

Interspecific breeding is often the most convenient way to achieve increased disease tolerance. Resistance may characterize one species that has evolved in the presence of high levels of a pathogen, while another lacks resistance, having evolved in the absence of the pathogen. Interspecific disease-resistance breeding has been successful in genera such as *Alnus,*

Fig. 8.2. A combination of genetic tolerance and predatory beetles effectively controls psyllid insect damage on four-month fodder regrowth of leucaena cultivar 'K636'.

Betula, Eucalyptus, Fraxinus, Leucaena, Pinus, Populus, and *Quercus.*
The transfer and use of disease-resistance genes in trees, not unlike that in
food-crop breeding, is only as effective as the epiphytotics under the breeder's control. Tolerance or resistance is characteristically available in two
genetic forms—monogenic, often gene-for-gene, and oligogenic. The latter
undergirds a general resistance, normally preferred for its evolutionary durability, which is not an immunity nor is it racially specific.

STRESS

Stress tolerance can be broadly defined to include tolerance of extreme soil
and environmental variations. Trees have often evolved with a curious mixture of broad and limited environmental tolerances. Broad tolerances are
required for adaptation in a single location to changes occurring through
seasons and years. Limited tolerance may be found to changes occurring
through space and to different soils or temperature regimes, including differences in soil pH and aluminum concentration or excessive heat or cold.
Species differences often characterize lowland from highland tropical tree
taxa, whose populations fail to overlap and thus to hybridize.

Wood yields of tree species are thus very site dependent. In the tropics,
they are often highly dependent on average annual temperature or soil characteristics such as the low pH of soils in Nakau, Indonesia or Iole, Hawaii
(Table 8.1).

Few breeders are more concerned about site adaptability than forest-tree
breeders. Trees are expected to grow almost anywhere, and it is more costly
to alter their natural environment to conform to growing requirements. The
differing targets of forest-tree improvement research dictate major differences in breeding approaches. One major target is the small tree farmer
seeking economic yields; another is the rangewide forester seeking forest
cover. Most breeding for agroforestry systems is directed to the first of these
targets, and in such cases one can expect that site amelioration and tree
breeding go hand-in-hand, as they must with food crops.

Widened adaptability has been achieved by tree breeders, notably in
improved tolerance to low temperatures and frosts and less clearly in im-

141

Table 8.1. Wood yield increments of fast-growing trees harvested after 2.5 years of growth.

Location	Avg. Temp. (°C)	Fresh Wood Yields (t/ha/yr)				
		LEUC	DIV4	MANG	AURI	MEAR
Nakau, Indonesia	28.6	11.9	12	27.5	—	—
Davao, Philippines	28.1	97.6	89.9	28.2	38.1	—
Waipio, Hawaii	24.2	30	23.2	16.3	12.8	45.2
Waimanalo, Hawaii	23.9	44.4	26.6	28.1	20.1	—
Molokai, Hawaii	23.4	81.9	46	16.7	18.3	81.6
Iole, Hawaii	20.1	2.8	24.9	37.8	12.2	107.4
Haleakala, Hawaii	19.3	2.7	18.4	17.1	14.8	65

Note: LEUC=*Leucaena leucocephala;* DIV4=*L. diversifolia* (4N), MANG=*Acacia mangium,* AURI=*A. auriculiformis,* MEAR=*A. mearnsii*

proved genetic tolerance to aluminum toxicity, drought, heat, and water-logging. Tolerance to diseases and pests has been exploited widely in a few tree species.

Breeding Systems

Methods for genetic advance differ depending on a plant's breeding system. These systems are often unknown for agroforestry species, but a useful generalization is that most diploids are outcrossing and many polyploids are selfing. Arabica coffee and *Leucaena leucocephala* are common examples of the latter. Self-incompatibility underlies the outcrossing in many species with perfect flowers, as it is characteristic of perennial plants in most plant families. Surveys of leguminous trees reveal that over 80% are self-incompatible. Genera like *Eucalyptus* include self-sterile trees as well as trees showing some inbreeding.

Plant exploration and seed collection should be restrained until information exists on the method of seed reproduction. High self-fertility allows the

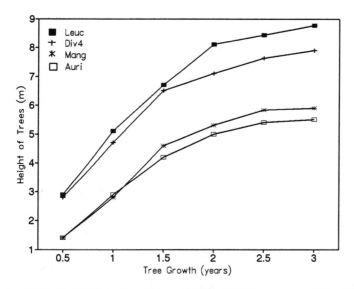

Fig. 8.3. Yields of selfed lines of two leucaenas (DIV4, LEUC) compared to yields of open-pollinated composites of two acacias (MANG, AURI) on agricultural soils in Hawaii.

breeder to identify superior accessions immediately, based on pure-line progenies from single trees. In this way, breeders of leucaena have made rapid genetic advance in two polyploid species (Fig. 8.3; see also Table 8.1). Outstanding selfed lines of L. *diversifolia* (DIV4) and L. *leucocephala* (LEUC) establish rapidly on good sites, outyielding competitive outcrossed acacias. Future breeding to exploit hybrid vigor or clonal uniformity will clearly provide major increments in yield for these species. Outcrossing species are normally seed-collected as broad-based provenances. Seed preservation on a single-tree basis is preferable here as well, because it permits studies of half-sib progenies and partial assessment of heritabilities.

The obligate outcrosser presents a major challenge in that genetic uniformity is virtually impossible in seed-derived progenies, even for simply inherited traits. Homogeneity is available in most trees, however, through vegetative propagation methods, including shoot and root propagation, grafting, marcotting, and tissue culture. In sustainable agricultural systems, uniformity is highly preferred for trees that yield fruits, is of less importance for timber trees, and is of little or no importance for trees that produce fodder, fuelwood, or green manure. A certain amount of heterogeneity is often preferred as a hedge against co-evolved pests and diseases.

The self-incompatible system of most angiosperms is based on a single multiple-allelic S locus. Pollen fails to grow down styles that carry the same S allele. Polyploidy can disrupt this incompatibility through a mechanism known as "competition interaction" that occurs in certain heterogenic diploid pollen from the polyploid anther. When such pollen fertilizes, the progeny automatically carries the same alleles and produces pollen that carries the same two alleles. Thus the self-fertility is perpetuated, and is disrupted only by rare outcrossing.

Provenance trials of outcrossing tree species often display no significant differences in yield. The high degree of heterogeneity of such species combines with the high land area requirements and variability to produce errors that can mask real varietal differences. Protracted data collection and multiple site testing can help clarify differences, but only at increased cost. The primary challenge remains to get trees evaluated in a suitable environment (with low coefficient of variability) where genetic differences in yield can be clearly discerned.

Polyploidy

Nearly half of our important cultivated plants are polyploids, and there is little reason to expect this to differ greatly for improved trees. Polyploidy characterizes many tree genera (Table 8.2), including 23 of the 77 *Acacia* species. Polyploids are more common among perennials than among annuals, and are often associated with increased size, making them of greater commercial interest.

Polyploid trees may be self-fertilizing while their diploid relatives remain self-sterile, as noted in the previous section. This facilitates approaches such as the production of seedless triploid hybrids using an isolated, cloned diploid parent as female.

The effects of polyploidy in plants are largely predictable:

• Larger cell size.
• Lower fertility.

Table 8.2. Polyploidy among nitrogen-fixing tree genera.

		Number of Species	
Genus	*Base 2N Numbers*	*Diploids*	*Polyploids*
Acacia	26	54	23
Albizia	26	8	1
Allocasuarina	22, 26	6	4
Alnus	28	12	9
Coriaria	20	0	4
Dichrostachys	26, 28	0	2
Leucaena	52, 56	10	4
Pterocarpus	22	2	4
Sesbania	12	3	3

• Genetic buffering.

• Slowed genetic segregations.

Many polyploid plants are amphiploids of two or more species; thus they often display expanded site adaptabilities over their diploid progenitors. An example created by the breeder is the leucaena hybrid LEUC x DIV4 (*L. leucocephala* x *L. diversifolia*) illustrated by data in Fig. 8.4. Hybrid yields in cool highlands greatly exceed those of the LEUC parent and equal those of the DIV4 parent, whereas they exceed those of both parents in the lowlands.

Genetic segregation of polyploids may be much less than that for diploids, demanding that breeders handle large populations with much patience. Importantly, polyploids can exploit multiple allelism (such as four different alleles at each locus in a tetraploid plant), which might assure better buffering against stress and damage caused by pests and diseases. Evidence for this buffering effect has been sought unsuccessfully. Segregations of a single locus synthesizing a tetrameric isozyme (such as catalase) in a polyploid like leucaena, can create more than 50 isozyme phenotypes.

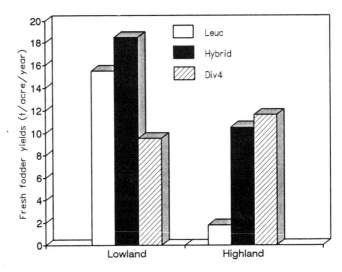

Fig. 8.4. Leucaena fodder yields over a two-year period in tropical highlands compared to yields in the lowlands.

Single genes are essentially useless in the polyploid plant. The search for useful radiation- or chemical-induced mutants is generally futile in polyploids. Searching for somatic mutations affecting ploidy, however, can be highly effective. Examples include polyploid chrysanthemum, bougainvillea, apple, banana, and breadfuit. Tree breeding programs should not be initiated without clear evidence concerning these chromosome relationships and breeding systems.

Species Hybridization

Species hybridization can be exploited directly or as a method for achieving new gene combinations or backcross conversions. The first objective is that of direct exploitation of the hybrids (see Fig. 8.3 above). Species hybrids have been exploited directly in genera such as *Acacia, Casuarina, Eucalyptus, Populus, Prosopis,* and *Salix.* Tree improvement is privileged to have the somewhat narrow objectives of improved yield and site adaptability. In contrast, the breeder of row crops is heavily constrained by human preferences of color, taste, odor, and chemistry. Improvement in trees may focus largely on yield, allowing for management options that reduce the need for genetic uniformity in form and appearance. Species hybrids of trees that produce unusable progeny or that are seedless can still be cloned and used directly, in contrast to those of annual species.

The exploitation of heterosis has only begun in multipurpose trees. Clones from hybrids of *Casuarina equisetifolia* x *C. junghuhniana* and *Casuarina equisetifolia* x *C. cunninghamiana* in Asia and *Eucalyptus* spp. internationally provide examples. Hybrids combining desired site adaptabilities can be expected. In addition, localized inbreeding can be very high among tree populations, and the vigor thus lost will be restored upon crossing. The capability of tree breeders to fix heterosis through clonal propagation drives the search for superior hybrids. The potential value of species hybridization appears to be very great for trees and shrubs in sustainable agriculture.

Fruit and seed production cause significant drain on carbon resources of many tree species. Tropical ornamental tree hybrids are normally selected for low seed production, often resulting from hybrid sterility. Notable examples are hybrid clones of banana, bougainvillea, *Cassia* spp., hibiscus, and frangipani that rarely set seeds. Leucaena breeding has also turned toward seedless triploid hybrids, in which improved wood and fodder productivity are evidently the result of better carbon partitioning (Fig. 8.1). Australian pine hybrids have also been selected for sterility or for dioecy, achieving seedlessness.

Ideotype and Yield

The much-debated concept of ideotype is that of a morpho-physiological model of a plant that should maximize yield. Applied to trees, models must consider optimal bole and crown structure, harvest index, and factors such as phenology, response to competition, and changes with age. Hopefully, such models will give a better prediction of yield and reduce field work. Nonetheless, trees find ways of maximizing yield that continually befuddle me, and I must agree with Simmonds (1985) that the best way to advance yield is to breed for it. Additionally, part of the joy of being a plant breeder is to assess and interpret yields in the field.

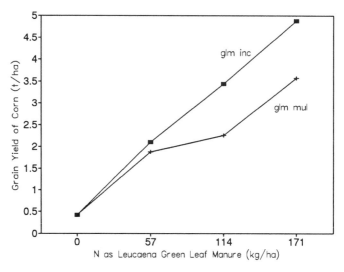

Fig. 8.5. Yield response to green leaf manure incorporated (glm inc) or mulched (glm mul) in a sustainable agricultural alley crop system with maize and leucaena.

Sustainable agriculture largely involves trees grown on short rotation cycles and harvested for fodder or wood. Fodder yields are estimated by direct weights, correcting for the proportion of the foliage that can be considered edible to the animal in question. In sustainable agriculture, trees and shrubs often serve to stabilize and renovate soils through processes such as nitrogen fixation, wind protection, soil stabilization, and altered ambient temperatures. Alley cropping systems often involve nitrogen-fixing shrubs in rows alternating with crops like maize. The yield of intercrop as influenced by leucaena green leaf manure, incorporated or applied as mulch (glm inc or glm mul), may respond linearly or nonlinearly to the trees' contribution (Fig. 8.5). Many variables influence yields and the effectiveness of green leaf manure, and few can be modeled accurately on computers in air-conditioned offices.

Wood yields normally involve measurements of tree height (h) and diameter ($d=2$ times the radius, r) at breast height. Wood yields are then most easily estimated by the volumetric formula for the volume of a cylinder

($\pi r^2 h$). Leucaena volumes are conveniently estimated quite accurately by $(d^2 h)/2$. The specific gravity of wood is then applied to give dry weights. Computers enable the calculation of much more elaborate formulas, often with only slight gains in accuracy but with variables that can be completely meaningless.

Suitability to Sustainable Agriculture

Most multipurpose trees are grown as part of agroforestry systems, and improved cultivars must fit the appropriate management system. Agronomists and foresters tend to debate *ad nauseam* the best species, systems, and ideotypes for such use, while neglecting to get the breeding started. Chuntanaparb and MacDicken (1991) emphasize that genetic advance for agroforestry systems requires that selection, or minimally evaluation, be conducted under agroforestry conditions. The interactions of tree with crop or animal or humans in sustainable agriculture are immensely complex; rarely are they entirely predictable or repeatable. Nonetheless, selection must involve judicious testing under the expected user conditions. A form of "shuttle breeding" is often appropriate, with one generation at a research station, a second under grower conditions.

Must the breeder evaluate only under the conditions of desired agroforestry management? Crop breeders have shown convincingly that high performance cultivars from the "Green Revolution" lead yield trials on poor as well as on good land. The few exceptions with high genotype-by-environment interactions (Souza et al., this volume, Chapter 10), seem to be related largely to good performance on poor sites (because of exceptional disease or pest resistance).

It is important that newly bred cultivars are suited to the management regimes to be applied. They cannot, however, be selected under the most limiting of farmer conditions. This is because genetic differences cannot be detected when "background noise" (experimental error variance, often expressed as the coefficient of variation) is too high (Goodman, this volume, Chapter 3). A simple example of this is shown in Fig. 8.6, with normal probability curves for segregants from a backcross segregating for a sin-

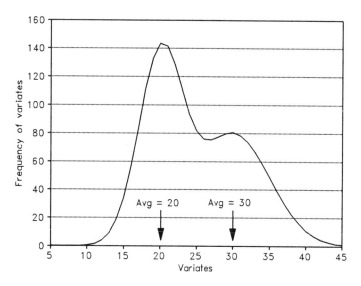

Fig 8.6. Normal probability distributions for two segregants differing by 10 units where standard deviation is 3 and 5 units.

gle gene (1000 data per each) where means and standard deviations differ slightly. Coefficients of variation increase only slightly (15 to 17%) in this example. The distributions are overlapping and statistical significance of the difference between them is lost as the coefficient of variation becomes larger.

Domestication

Genetic improvement is a continuum involving natural selection and radiative adaptation, domestication, and the growth of natural or artificial progenies in what is normally thought of as plant breeding. Many multipurpose trees are in the first of these three phases, and only a few have been fully domesticated. Domestication of trees often occurs outside the region of origin and rarely involves a thorough sample of germplasm—consider Monterey pine, monkeypod, Australian pine and *Eucalyptus* spp. One genotype of leucaena circumnavigated the globe 400 years ago in this phase of domestication. Only in the 1960s were serious efforts made to enlarge its germplasm base.

Domestication should be a corollary to all tree breeding programs that involve lesser known species or provenances. Their site adaptabilities are best tested as an adjunct to replicated trials of adapted materials. The augmented block design of Federer (1961) is a useful experimental tool when such provenances (entered unreplicated over sites) are not too dissimilar from adapted lines (entered replicated over sites).

Strategies for Genetic Improvement

Strategies for genetic improvement differ little from herbaceous to woody plants, especially when life cycles are short. These strategies attempt to maximize capture of the two major types of genotypic variance, additive and nonadditive. General combining ability (GCA) is a reflection of additive variance, the main cause of the resemblance of related plants and the basic tool of the plant breeder. Specific combining ability (SCA) reflects nonadditive genetic variance, related to allelic and gene interactions.

Heritability expresses the ratio of total genetic variance to total phenotypic variance. When heritabilities are high, rapid genetic advance can be made with simple mass selection. Further advance is ensured by selection from pedigreed half-sib (no control of male) or full-sib (controlled male) families. Recurrent selection methods involve the repetition of generations of selection of families or plants, customarily leading to improved general combining ability with 3–6% progress per generation.

A basic issue in tree improvement is how broad the base for genetic improvement needs to be. In general, tropical trees show as great variation within (outcrossed) provenances as they do between provenances. Thus, individual tree progenies are the primary starting point for much improvement. Individual tree progenies are grown in yield trials for selection and can then be rogued to leave outstanding trees as seed parents for the next cycle of selection. Provenance variations are observed occasionally, generally where species range widely in latitude or altitude.

Simple recurrent mass selection is the breeding strategy that maximizes progress for fast-growing trees that are highly outcrossing. This method exploits GCA at relatively low cost when large heterogeneous populations are grown under conditions that minimize environmental variance. Good progress has been recorded for cold tolerance by this method in leucaena and *Eucalyptus* spp.

Strategies to Accelerate Genetic Gains

Genetic gains reflect a response to selection. They can be calculated from the formula:

$$G = i * h^2 * \sigma_p$$

In this formula, genetic gain (G) is given as the response to selection intensity (i), for a trait of heritability (h^2), where available phenotypic variability has a standard deviation of σ_p. Selection intensity values are tabular; 20% selection in a population of 100 individuals gives $i=1.4$, whereas 1% selection gives $i=2.5$. The "selection differential" (s) between the mean of a selected group and that of its source population equals $i*\sigma_p$. Thus, genetic gains from selection can be viewed simply as $s*h^2$. These concepts are discussed thoroughly by Falconer (1989), Namkoong (1979), Wright (1976), and Zobel and Talbert (1984).

Tree experiments must be designed to optimize expression of the genes of interest, thus maximizing h^2 and σ_p. As an example, consider a population whose tree heights average 12 m with σ_p equal to 1 m. Assuming $h^2=0.71$, and $i=1.4$ (20% selected in a population of 100 trees), the expected genetic advance through selection would be about 1 m per cycle of selection.

In a thorough review of the subject of tree improvement, Matheson (1990) stresses that the relationship between i and the fraction of a population selected is exponential. Thus, where 40 trees are adequate to serve as a basis for selection advance where $i=2$, it requires a base of 32,000 trees to make the same genetic gain where $i=4$. High levels of i can therefore be difficult or impossible with trees. A basic problem is that mature trees require large land areas, and environmental heterogeneity (and variance) increases with land area, thus reducing h^2 (which is the ratio of genetic: genetic + environmental variation). In some instances, selection can be made using less land area at the seedling stage; thus, major genetic gains for thornlessness in *Prosopis pallida* were made by selecting less than 1% of plants from populations of 4000.

Genetic gains are similarly complicated where two or more traits are selected simultaneously. Selection indexes permit the simultaneous selection of several traits, but selection pressure must remain low unless population size can be very large.

An important and unique strategy for accelerating genetic gain is available to most tree breeders—the use of clonal offspring. Small clonal progenies planted under differing environmental conditions quickly elucidate relationships of genetic and environmental components of variance. The ability to clone also allows the capture of SCA. Heterosis and inbreeding depression are part of this phenomenon and vary depending on the breeding system of the species studied. Obligate outcrossers have high inbreeding depression. Self-pollinators show little or no inbreeding depression, and SCA can probably best be harnessed through species hybridization. The costs of selection for SCA are high, involving controlled pollination and genetic testing, and are a bit premature for many multipurpose trees.

The Future

Plant breeding is one of the most personally and publicly rewarding sciences, the impact of which has been felt on the economies of every nation and people (Willan, 1988). It is also a simple science in which progress requires as much perseverance as skill. Despite the incredible scholarship of modern molecular and statistical genetics, plant breeders still rely effectively on broad germplasm bases; good experiment stations; and the breeder's "eye," experience, and love of plants. We do great injustice to tropical foresters to limit their training to air-conditioned laboratories.

Great plant breeding progress has been made by dedicated individuals working as amateurs. Most of the noted cultivars of tree fruits arose through grower interest and intellect. Major progress in multipurpose tree improvement can be expected in the future from these "barefoot breeders," as they are denoted by Chuntanaparb and MacDicken (1991), a term derived from the community-based "barefoot doctors" of China. There should be official recognition and compensation for their efforts, and society should encourage appropriate forums for the sharing of their technologies and germplasm. Anyone familiar with the incredible productivity of California's "barefoot breeder" Luther Burbank cannot help but explore the opportunities.

Genetic improvement of multipurpose trees for sustainable agroforestry systems in the future, however, probably depends as much on political will as on breeders' skill. Plant breeding requires time, no matter how short the plant's life cycle. Regrettably, the research life of tree breeders in developing countries is often less than that of the seed-to-seed cycles of the trees themselves. Long-term commitments of funds and institutional programs are essential if tree improvement is to be much more than a topic for chapters such as this.

References

Brewbaker, J.L. 1987. Leucaena: A genus of multipurpose trees for tropical agroforestry. p.289–323. In H.A. Steppler and P.K.R. Nair (ed.) Agroforestry: A decade of development. International Center for Research on Agroforestry, Nairobi, Kenya.

Chuntanaparb, L., and K.G. MacDicken. 1991. Tree selection and improvement for agroforestry. p. In N. Glover and N. Adams (ed.) Tree improvement for multipurpose species. Winrock International, Arlington, VA.

Falconer, D.S. 1989. Introduction to quantitative genetics. 3rd ed. Longman, New York, NY.

Federer, W.T. 1961. Augmented designs with one-way elimination of heterogeneity. Biometrics 17:447–473.

Matheson, A.C. 1990. Breeding strategies for multipurpose trees. p.67–99. In N. Glover and N. Adams (ed.) Tree improvement for multipurpose species. Winrock International, Arlington, VA.

Namkoong, G. 1979. Introduction to quantitative genetics in forestry. USDA Forest Service Bull. 1488. U.S. Govt. Print. Office, Washington, DC.

National Academy of Sciences. 1984. Leucaena: Promising forage and tree crop for the tropics. 2nd ed. National Academy Press, Washington, DC.

Simmonds, N.W. 1985. Principles of Crop Improvement. Longman, New York, NY.

Willan, R.L. 1988. Economic returns from tree improvement in tropical and sub-tropical conditions. Danish International Development Agency Tech. Note 36, Danish International Development Agency, Humlebaek, Denmark.

Wright, J.W. 1976. Introduction to forest genetics. Academic Press, New York, NY.

Zobel, B.J., and J.T. Talbert. 1984. Applied forest tree improvement. Wiley, New York, NY.

Additional Reading

Brewbaker, J.L. 1963. Agricultural genetics. Prentice-Hall, Englewood Cliffs, NJ.

Brewbaker, J.L. 1987. Significant nitrogen fixing trees in agroforestry systems. p.31–46. In H.L. Gholz (ed.) Agroforestry: realities, possibilities, and potentials. M. Nijhoff and Dr. W. Junk, Dordrecht, Netherlands.

Brewbaker, J.L., and C.T. Sorensson. 1990. New tree crops from interspecific leucaena hybrids. p.283–289. In J. Janick and J. Simon (ed.) Advances in New Crops. Timber Press, Portland, OR.

Briscoe, C.B. 1990. Field trials manual for multipurpose tree species. 2nd ed. Winrock International, Arlington, VA.

Burley, J. 1980. Choice of tree species and possibility of genetic improvement for smallholder and community forests. Commonwealth Forestry Rev. 59(3):311–325.

Burley, J., and P. von Carlowitz (ed.). 1984. Multipurpose tree germplasm. International Center for Research on Agroforestry, Nairobi, Kenya.

Burley, J., and P.J. Wood. 1976. A manual on species and provenance research with special reference to the tropics. Tropical Forestry Paper 10. Oxford Forestry Inst., Oxford, England.

Gibson, G.L., A.R. Griffin, and A.C. Matheson. 1990. Breeding tropical trees: Population structure and genetic improvement strategies in clonal and seedling forestry. Winrock International, Arlington, VA.

Glover, N., and N. Adams (ed.). 1990. Tree improvement for multipurpose species. Winrock International, Arlington, VA.

National Academy of Sciences. 1980. Firewood crops: Shrub and tree species for energy production. Vol. 1. National Academy Press, Washington, DC.

National Academy of Sciences. 1983. Firewood crops: Shrub and tree species for energy production. Vol. 2. National Academy Press, Washington, DC.

Palmberg, C. 1986. Selection and genetic improvement of indigenous and exotic multipurpose tree species for dry zones. Agroforestry Systems 4:121–127.

9

Contributions of Biotechnology
to Crop Improvement

Susan R. McCouch, Pam Ronald, and Molly M. Kyle

Biotechnology refers to an array of laboratory methods developed over the last forty years that are based on advances in cellular and molecular biology. Some techniques, such as embryo rescue, micropropagation, and anther culture, have been applied by plant breeders in crop improvement programs for decades. A number of others, such as molecular marker technology, gene construction, and transformation, are only in the initial stages of integration into plant breeding. In most cases, it is the breeding process rather than the breeding product that is radically affected by these new tools or biotechnologies. Improved cultivars generated with the assistance of the latest cellular and molecular techniques are still packaged in the traditional way—in the form of a seed.

There are two principal ways in which cellular and molecular technologies affect the breeding process: by increasing the efficiency of selection and by generating new forms of genetic variation. Whether or not these tools offer practical and cost-effective alternatives to traditional breeding approaches inevitably depends on the objectives, the crop, and the resources available to the program.

The aim of this chapter is to point out existing opportunities for using biotechnology to achieve goals of enhanced stability and sustainability of agricultural systems. In order to be sustainable, an agricultural system must be viable over decades or centuries. It must contribute to the welfare of humankind while safeguarding the quality of the environment.

Different physical and social environments sustain different types of production systems, crops or cultivars, cultivation practices, and marketing and distribution networks. Though there are many ways to approach the challenge of sustainability, a factor that increasingly threatens any agricultural production system is human overpopulation. Sustainable agriculture must be self-regulating in terms of food production and distribution potential in relation to human potential. If premised on a fundamental respect for the long-term requirements and limitations of both biological and social systems rather than on short-term gains, innovations in the area of biotechnology can contribute to achieving an acceptable quality of life for people on this planet.

We have deliberately limited our discussion to areas of biotechnology research where there has been enough progress to assess how specific methods may contribute to increasing the stability and sustainability of agricultural production systems. We will discuss how molecular markers, tissue culture, and transformation technology are useful in evaluating, using, and increasing crop genetic diversity and in reducing dependence on external factors such as purchased agricultural inputs. Many of our examples will be in the area of pest resistance and will be drawn from current applications in rice improvement. We have chosen to emphasize rice because of the importance of this crop to human populations and because of the concerted international effort to develop and integrate biotechnological approaches into rice improvement.

At the same time, it is our intention to identify limitations of current biotechnologies and to critically evaluate how future developments may alter current realities. In light of the rapid evolution of cellular and molecular techniques, it is important to keep in mind that possibly the most exciting and influential opportunities for applying biotechnology to crop improvement for sustainable agriculture have yet to be imagined.

Evaluation and Use of Genetic Diversity

Genetic diversity is the essential ingredient of plant breeding. It is the result of natural evolution and is an important component of sustainable, ecologically balanced biological systems. Diversity ensures the long-term adaptability and survival of a population. Where individual fitness varies depending upon the climatic, edaphic, and biotic components of the immediate environment, collections of individuals with varying fitness attributes contribute to the genetic diversity of a population. Complex, natural ecosystems tend to be genetically complex, whereas agricultural systems represent a simplification of that complexity. In the following section, three very different types of biotechnological tools will be defined and discussed in terms of their contribution to the evaluation and enhanced use of genetic diversity in crop plants.

Over time, crop improvement for modern agriculture has contributed to the reduction of genetic diversity because the most productive and profitable genotypes are often grown in monoculture over large areas. For example, the rice variety 'IR36' was grown on an estimated 10.5 million ha in 1984 (Khush, 1984). Such widespread popularity of a single inbred variety meant that a small number of genes was deployed over a very large geographic area. Though this was the result of a success story in plant breeding, it was accompanied by consequences unforeseen before the Green Revolution, such as cycles of pest epidemics.

Widespread planting of genetically uniform crops may contribute to the long-term destabilization of yields, despite short-term gains in production, by imposing unacceptable levels of selection on pest populations. This selection pressure can have devastating consequences in epidemic-prone environments. Two of the best known examples of such a phenomenon are the southern corn leaf blight epidemic that occurred in the United States in 1971 (National Academy of Sciences, 1972) and the breakdown of brown planthopper resistance in rice in Asia during the 1970s (Khush, 1984). Equally prone to the boom-and-bust cycle is that of the rice blast fungus, where rapid shifts in the races of this pathogen have historically outstripped breeders' efforts to generate stably resistant varieties (Chuke and Bonman, 1988). Plant breeders continually seek useful new forms of genetic diversity and new ways of using that diversity in an attempt to stabilize yields in the face of new stresses. Yield stability is an important component of sustainable agriculture and may be achievable if increased levels of useful genetic diversity can be incorporated into highly productive cropping systems.

Both traditional and novel forms of genetic diversity are available to plant breeders. The vast array of naturally occurring diversity found within most species offers seemingly endless opportunities to improve crops. Both *in situ* and *ex situ* germplasm collections strive to make this variation available for breeding purposes. One role of biotechnology is to make germplasm more readily accessible and enable breeders to manage, evaluate, and use germplasm collections in more efficient ways than ever before. Novel forms of variation created through genetic engineering are now also available thanks to transformation technology. Deployed appropriately, the enhanced use of available genetic diversity can help stabilize agricultural yields, thereby contributing to sustainability. Technologies that will be discussed are molecular markers, cell and tissue culture, and transformation.

MOLECULAR MARKERS

Two types of DNA markers, restriction fragment length polymorphisms (RFLPs) and randomly amplified polymorphic DNA (RAPDs), have served as the backbone of genome mapping efforts over the past decade. These markers are being modified and adapted to serve as diagnostic tools in human medicine as well as plant and animal agriculture. DNA markers provide a way to evaluate genetic diversity and detect the presence or absence of target genes at the genotypic level. New generations of molecular markers increase the attractiveness and utility of this technology to plant breeders by making it progressively less expensive and simpler to use.

RFLP technology involves the use of cloned DNA fragments (the markers) that are used as probes to monitor genetic variation. Genetic differences between organisms are the result of mutations, or changes in DNA, which accumulate over time. This DNA-based marker technology permits the detection of these changes, which are observed as size variation among DNA restriction enzyme fragments (Botstein et al., 1980).

A newer form of molecular marker, RAPD, is composed of segments of DNA that are amplified *de novo* in the genotypes of interest and visualized as bands of varying molecular weight on an agarose gel. In an enzymatically catalyzed reaction known as the polymerase chain reaction (PCR), specific primer sequences (short oligonucleotides) are incubated along with DNA polymerase (the enzyme that allows the synthesis of a complementary second strand of DNA from a single-stranded template) and DNA from individual genotypes. If the primers bind to regions of the genome in the correct orientation and within a short distance (less than 5 kilobases) from one another, a discrete fragment of DNA will be amplified during the course of the reaction. Thus, differences in nucleotide sequence, therefore lengths between points at which primers anneal, determine the size of the fragments that are amplified during an RAPD reaction (Martin et al., 1990; Williams et al., 1991).

Both RFLP and RAPD markers represent genetic loci that can be mapped and used to locate genes of agronomic importance via linkage analysis. RFLPs were first identified as useful genetic markers by Botstein et al. (1980). Since that time, genetic maps based on RFLPs have been constructed for plant species of agricultural importance including tomato, maize, lettuce, rice, potato, barley, pepper, and soybean (Bernatzky and Tanksley, 1986; Bonierbale et al., 1988; Helentjaris, 1987; Heun et al., 1991; Landry et al., 1987; McCouch et al., 1988; Tanksley et al., 1988; Tingey et al., 1989). Maps based on molecular markers are also available for a number of bacteria and fungi of importance to plant improvement (Hulbert et al., 1988; Leung and Taga, 1988).

MOLECULAR MARKERS IN GERMPLASM EVALUATION

Germplasm collections constitute attempts to preserve genetic diversity as environmental degradation threatens ecosystems in which the germplasm evolved and currently exists (Chang, 1984; Oka, 1991). Loss of genetic diversity has gained worldwide attention as concerns over the state of the earth's environment increase and an international Biodiversity Conservation Strategy is being organized (Biodiversity Conservation Strategy Update, 1991). The potential value of germplasm collections for plant improvement programs cannot be realized unless they are appropriately classified, evaluated, and maintained.

Molecular markers, including isozymes and mitochondrial and chloroplast restriction analysis can be useful in assessing genetic variation in natural populations and germplasm collections. In rice, their use has been effectively demonstrated over the past 15 years by many workers (Barbier et al., 1991; Dally and Second, 1990; Endo and Morishma, 1983; Glaszmann, 1987; Ishii et al., 1988; Ishii et al., 1986; Second, 1985; Second, 1982; Wang et al., 1992; Wang and Tanksley, 1989). Markers are also being used to evaluate world collections of maize, *Brassica* spp., bean, tomato, potato, and manioc, among other crops.

Despite the tremendous potential of molecular markers, many questions still need to be answered before the benefits of molecular marker technology can be fully realized. For example, an absence of diversity at the molecular level does not necessarily correlate with an absence of genetic diversity useful to plant breeders. Several important crops (e.g., tomato, Miller and Tanksley, 1990; melon, Shattuck-Eidens et al., 1990; and wheat, Kam-Morgan et al., 1989; Liu et al., 1990), show very little polymorphism assessed with restriction endonucleases but possess sufficient morphological variability to permit very dramatic advances by traditional plant breeding methods. Maize, on the other hand, embodies sufficient RFLP and RAPD polymorphism for these molecular markers to be useful even in distinguishing between related inbred lines (Evola et al., 1986; Helentjaris et al., 1985).

One of the great advantages of using RAPD markers is that DNA may be obtained from a very small sample of material from virtually any plant part, and even from herbarium specimens. Thus, plants with very few poor quality seeds, or those that may be very difficult to classify phenotypically outside of their native environments, are readily evaluated using molecular approaches. Molecular markers also offer a reliable assessment of the amount of genetic variation within as well as between germplasm accessions. Such information can be critical in deciding how to amplify a seedlot or maintain the genetic integrity of an accession.

A consequence of molecular evaluation is the identification of clusters of germplasm that embody unusually large amounts of allelic variation. These clusters frequently correspond to particular geographic or ecological zones that may be targeted for further collection. Judicious use of molecular information may greatly increase the efficiency with which the very scarce resources for germplasm maintenance and collection are expended.

Molecular markers enhance germplasm use by identifying and locating genes controlling characters of interest to breeders, such as resistance to diseases, insects, and stresses (McCouch et al., 1991; Ottaviano et al., 1991; Ronald et al., 1992; Walbot and Gallie, 1991; Yu et al., 1991). They are also being used to study the genetic basis of heterosis, yield, and quality characteristics (Fatokun et al., 1992; Godshalk et al., 1990; Li et al., 1991; Paterson et al., 1988; Stuber et al., 1992). Once genes of interest have been located, or "tagged," markers linked to these genes are used to identify rare recombinants, break undesirable linkages, and assist in efficiently transferring genes of interest from either cultivated or wild species into different genetic backgrounds (Murray et al., 1988; de Vicente, 1991; Paterson et al., 1991). In the past, rice breeding was largely dependent on the pedigree method, repeatedly using well-adapted elite lines as parents in crossing programs (Dilday, 1990; Hargrove et al., 1988). The use of agronomically poor parents that possessed traits of interest was uncommon because of the additional time required to purge subsequent generations of the undesirable traits introduced along with those of interest (Goodman, this volume, Chapter 3). Molecular markers help circumvent this problem, permitting more rapid transfer of genes from unadapted germplasm to elite lines.

TISSUE AND CELL CULTURE

Tissue and cell culture are *in vitro* methods useful in manipulating cells or organs, propagating disease-free plants, and regenerating whole plants from a variety of plant parts. Included in this broad area are techniques such as embryo rescue, meristem culture, anther or pollen culture, protoplast fusion, and protoplast culture.

Tissue culture techniques such as embryo rescue have been used for over 70 years (Buckner and Kastle, 1917) and have had an impact on plant improvement (Ivanovskaya, 1946; Iyer and Govila, 1964; Sitch et al., 1989; Yeung et al.,1981). Embryo rescue involves the *in vitro* culturing of embryos. The technique is useful in germplasm management and is frequently employed following interspecific hybridization for embryos that would not complete development without intervention. Whole plants regenerated from rescued embryos are typically backcrossed to the more agronomically desirable parent to recover progeny carrying a high percentage of desirable traits while retaining the useful gene(s) from the other parent, usually a wild relative (Brar et al., 1991).

Another technique being profitably employed in plant improvement programs and commercial production is *in vitro* propagation of whole plants from plant parts (such as undifferentiated callus tissue or root and shoot meristems). *In vitro* propagation is especially useful in the production of certified, disease-free plants in vegetatively propagated tree fruits, strawberry, asparagus, and potato (Plucknett et al., 1990).

Anther or pollen culture can be employed to generate haploid callus containing only the male set of chromosomes (Bhojwani and Razdan, 1983). When haploid callus cells double their chromosomal complement, they can be regenerated to produce homozygous plants known as *doubled haploids*. Production of these instant inbreds is accomplished in one generation from F_1 anthers. Since traditional breeding methods of producing relatively homozygous inbred lines take six to eight generations, the production of doubled haploids represents a substantial improvement in breeding efficiency. This is also an efficient way to fix desirable characters and to identify recessive traits. Difficulties encountered in regenerating doubled haploids restrict their use in certain plant species or genotypes.

Protoplast fusion is a technique in which the entire cell contents (rather than just the nuclei) of two plants are merged. The widespread use of protoplast fusion is limited by difficulties in the regeneration of plants following fusion, chromosomal incompatibility, and the lack of control over transfer of specific genes or chromosomes. One promising example for the application of protoplast fusion to cultivar improvement has been demonstrated in citrus (Grosser and Gmitter, 1991). Fertile intergeneric somatic hybrids between citrus and sexually incompatible related species were produced incorporating novel traits into potentially useful rootstock material.

Anther culture technology may also contribute to sustainable agriculture, creating new forms of genetic variation. Crosses between *indica* and *japonica* subspecies of rice are difficult to manage because of a sterility barrier that commonly appears in the F_1 or F_2 generation and becomes increasingly evident in subsequent generations of inbreeding. In addition to sterility, allele frequencies may be skewed by the growing environment because *indica* and *japonica* types are adapted to different ecologies and climates (Oka, 1991). Through the use of tissue culture, however, doubled haploids with the expected 50% *indica* and 50% *japonica* alleles have been generated (Guiderdoni et al., 1989).

Some doubled haploids are performing exceptionally well. 'Texmont' is an example. It was the first U.S. rice variety produced via anther culture and was released in 1990 by the Texas A&M Rice Improvement Association (A. McClung, personal communication). Many other anther culture–derived lines have been used as parents in traditional rice breeding programs throughout the world. In fact, production of anther culture–derived doubled haploids is becoming routine in rice breeding.

TRANSFORMATION

One of the most striking examples of the potential for biotechnology to increase genetic diversity is through transformation, which involves the insertion of naked DNA directly into a cell. Traditional plant breeding and marker-assisted selection depend on the availability of useful genes in the crop of interest or closely related species. In many cases, the reservoir of genetic variability in cultivated species provides ample opportunity to select characters of interest using traditional methods. For some crops or characters, however, existing genetic variation is difficult to manipulate or inadequate to achieve the desired objectives. Plant transformation technologies can circumvent these limitations through direct DNA transfer across species barriers to introduce novel forms of genetic variation. Since the first plant transformation was carried out on tobacco in 1984 (Paszkowski et al., 1984) nearly 50 additional crops have been transformed (Fraley, 1992).

Agrobacterium-mediated transformation is the most efficient and straightforward transformation method at this time and is useful in many dicotyledonous species (Gasser and Fraley, 1990). Other DNA delivery systems have tremendous potential for transforming monocots and plants that are not amenable to *Agrobacterium*-mediated transformation. Other methods include PEG-mediated uptake of DNA through protoplast membranes, electroporation, microinjection, particle bombardment, and alternative vector transmission (Cao et al., 1991; Uchimiya et al., 1989). Most of these techniques require *in vitro* regeneration of plants following cell or tissue transformation. A large percentage of the problems encountered in transformation are the result of difficulties in recovering fertile progeny from transformed plant parts. Regeneration is easier in some species than others and within a species some genotypes are more responsive to culturing techniques than others. Among the cereal crops only rice and maize have been successfully transformed and regenerated (Gordon-Kamm et al., 1990; Shimamoto et al., 1989). Within *Oryza sativa, japonica* varieties can be regenerated with much greater efficiency than *indica* varieties (Hodges et al., 1991a).

Meaningful opportunities to use transformation in crop improvement depend on the availability of cloned genes of interest (Toenniessen, 1991). To date, relatively few agronomically important genes have been isolated, but this is an area where rapid progress is being made. For effective transformation, not only must the sequence encoding the gene product itself be cloned but the appropriate regulatory sequences must be isolated as well. Cloned genes and appropriate regulatory sequences must then be assembled on a recombinant DNA molecule (Walbot and Gallie, 1991). Once such a construct has been prepared, it is introduced into a recipient cell to become integrated into the native chromosomal complement. If it integrates, it will be inherited in a Mendelian fashion. If it does not integrate, the gene may still be effectively transcribed and translated, resulting in transient gene expression. Some unpredictable consequences may be observed following successful transformation, especially when several copies of an introduced gene become integrated into the host genome (Deroles and Gardner, 1988; Hodges et al., 1991a; Matzke et al., 1989).

Molecular marker, tissue culture, and transformation techniques are being integrated into existing plant breeding programs, allowing researchers to access, transfer, and combine genes at a rate of precision previously not possible. Tissue culture techniques such as embryo rescue offer a way of incorporating useful genes from wild species into cultivated crops. Genetic engineering and transformation allow useful genes from even unrelated organisms to be incorporated into the gene pool of productive cultivars. Molecular marker technology is useful in assessing genetic variation, allows rapid confirmation of the presence or absence of genes of interest, and facilitates selection in a breeding program. Though the research and development costs of many of these techniques have been high, the rapid development of less expensive and simpler methods is making many of the tools of biotechnology into useful breeding aids suitable for broad application in germplasm characterization and selection. As these techniques become more accessible, they offer breeders new ways of expanding the available gene pool for the improvement of cultivated crops, increasing the opportunities to stabilize yields and tailor varieties to specific environments.

Reduction of Agricultural Inputs

Over the past 40 years, worldwide deployment of modern cultivars has led to dramatic increases in yield for many crops, especially staple crops such as rice, wheat, and maize. These yield increases, however, typically relied on increased amounts of fertilizers, pesticides, and water (Whitten and Oakeshott, 1990). The evolution of pesticide-resistant pests and weeds, increased dependence of farmers on agrochemicals, and the disruption of fragile agroecosystems were related consequences. Clearly, biotechnological innovations reducing dependence on agrochemicals would be of great advantage for sustainable cropping systems. In this section we give examples of biotechnological strategies directed at reducing the use of pesticides.

INSECT RESISTANCE

Bacillus thuringiensis (Bt) is a bacterium capable of producing a toxin-containing crystal protein that is highly active against several insect orders, including Lepidopteran larvae (Lambert and Peferoen, 1992), Coleoptera (Hofte et al., 1987; McPherson et al., 1988), and Diptera (Goldberg and

Margalif, 1977), as well as nematodes (Meadows et al., 1990). The crystal protein has been widely deployed for a number of years in various commercial biopesticide formulations. Major limitations to its use in this way have been its short residual activity and more importantly, rapid evolution of resistance in target pests (Tabashnik et al., 1990).

Bt was a logical choice in early attempts to create insect-resistant crops with gene insertion technology, despite concerns about insect resistance to the toxins. The prokaryotic origin of the gene greatly simplified cloning efforts. Numerous versions of the Bt toxin have been isolated and in some cases the structural sequence has been altered, allowing manipulation of toxin expression in plants (Perlak et al., 1991).

Toxin expression is manipulated by using a promoter that allows gene expression and toxin production only at specific developmental stages in plant growth (such as, when crop yield is most vulnerable to the damage inflicted by Lepidopteran larvae), in specific tissues (such as those most affecting yield), or in response to specific inducers (such as wounding or inert substances that could be sprayed on the crop with no harmful environmental effects but would serve to trigger toxin production). All of these approaches minimize pest contact with the toxic protein, thus minimizing selection pressure for toxin resistance in pest populations (Gould, 1988). Though feeding by the larvae would not be entirely restricted using such a strategy, economically significant damage could be avoided. Judicious deployment of transformed crop plants containing such carefully engineered forms of resistance will initiate a new era in integrated pest management.

VIRUS RESISTANCE

One of the most successful strategies for engineering resistance to plant viruses is coat protein–mediated resistance. This involves cloning the coat protein sequence of a virus and using it to transform a host plant (Beachy et al., 1990; Powell-Abel et al., 1986). Expression of the coat protein in the host plant interrupts the virus infection cycle and reduces symptom expression to varying degrees. The expression of viral coat protein in the plant endows the plant with a form of cross-protection against identical or related viruses. The effectiveness of coat protein–mediated resistance typically depends on the viral titer as well as the time of infection and environmental conditions.

An example of current research aimed at engineering resistance based on the viral coat protein model involves rice tungro disease (RTD). RTD is a complex disease involving both a bacilliform, double-stranded DNA virus (RTBV) and a spherical RNA virus particle (RTSV), as well as an insect vector, the green leafhopper (Glh). The disease can be devastating to rice production, but little was known about its etiology before it became a target of research at the molecular level. An infectious clone of RTBV has been isolated and sequenced, the coat protein sequence has been identified, and a phloem-specific tungro virus promoter has been used to successfully transform rice with a reporter gene (*gusA*) (Beachy et al., 1991; Hodges et al., 1991b; Hull et al., 1991). The objective is to engineer a rice plant that will produce the viral coat protein predominantly in phloem cells (the Glh is predominantly a phloem feeder) so that inhibition of the infection cycle will be effective as soon as the virus enters the plant. Such a targeted delivery system aims to minimize the amount of coat protein that the plant must produce and maximize the speed of initiation of coat protein production and the effectiveness of disease inhibition.

Engineered coat protein–mediated virus resistance represents a source of genetic variation that can be readily incorporated into a traditional breeding program once transformed plants carrying the viral genes are available. Whether the widespread use of genes from such novel sources will be any more effective in protecting plants from pests and diseases in the long term is still an open question. In the United States, the preliminary testing of tomato plants transformed with TMV coat protein genes and of Russet Burbank potato plants expressing the coat protein genes of potato virus X and potato virus Y has produced promising results (Kaniewski et al., 1990; Nelson et al., 1988), but field testing is still very restricted.

In the case of the tungro virus disease, native host gene resistance has been difficult to identify. Resistance to the Glh vector was identified as early as 1971 (Athwal et al., 1971). It was difficult to distinguish virus resistance from Glh resistance because infection with the virus required vector transmission. With the aid of infectious clones of the virus that have been isolated by Hull et al. (1991), rice germplasm can now be screened for native genes conferring virus resistance. The strategy of engineering coat protein–derived virus resistance and using it in combination with native genes for resistance to RTD promises new opportunities for breeding rice varieties with durable resistance to this complex disease. Molecular markers will be invaluable in selecting plant material carrying several different forms of resistance, each of which confer similar phenotypes.

In addition to coat protein genes, other viral sequences can confer resistance. In at least a few cases, cloned portions of genes involved in viral replication can confer extremely high levels of resistance, though a high degree of specificity may limit the breadth of applicability across different isolates of a particular virus (Golemboski et al., 1990). Other sequences being evaluated for use include viral satellite and other replication-dependent small molecules capable of attenuating disease symptoms (Collmer and Howell, 1992; Tien and Wu, 1991). A major concern with these strategies revolves around the fact that these molecules can have opposite effects in different host species; thus widespread deployment to protect one crop could cause increased losses in another if transferred by a viral vector.

NATIVE PLANT RESISTANCE GENES

A different strategy for engineering pest resistance involves cloning and inserting naturally existing host resistance genes. Several plant defense-related genes, including proteinase inhibitors, hydrolytic enzymes such as chitinase and glucanase, and genes involved in the phytoalexin biosynthetic pathway have been cloned and extensively characterized (Dixon and Lamb, 1990; Hilder et al., 1987; Nishizawa and Hibi, 1991; Zhu and Lamb, 1991). Unlike the single plant resistance genes traditionally used by breeders, however, these genes have not been shown to be the sole determinants of resistance. They may, in fact, be part of a cascade of events downstream from the initial recognition event triggered by pathogen infection. Although considerable efforts have been targeted on cloning single genes known to confer a useful level of host resistance in the field (Ganal et al., 1990), there are few reports of success in this area to date (Johal and Briggs, 1991).

A number of factors contribute to the difficulty in cloning these eukaryotic genes, including the fact that plant genomes are generally 10^1 to 10^3 times larger than bacterial genomes and 10^4 to 10^6 times larger than viral genomes, and generally the resistance gene products are either not known or are expressed at very high levels. With the advent of new molecular techniques it is likely that disease-resistance genes will be cloned from several crops in the coming years. As the first of these genes are cloned and their gene products studied, new strategies for creating or enhancing native forms of resistance are emerging. The potential to manipulate the level, timing, or location of expression of resistance genes increases the interest in isolating these genes. It is also possible that once cloned, a resistance gene from one plant can be effectively transferred into an unrelated crop species to provide protection against related pathogens. As more cloned host resistance genes become available, it will be possible to evaluate the extent to which they can function in heterologous genetic backgrounds.

Some of the most exciting and innovative strategies for manipulating plant defenses involve the manipulation of plant processes that are not typically involved in the infection cycle. Examples include directed RNAse activity or RNA-catalyzed site-specific cleavage by ribozymes (Cotten, 1990; Haseloff and Gerlach, 1988; Izant, 1989; Rossi and Sarver, 1990). These applications of biotechnology are emerging directly out of basic research laboratories and often exemplify the situation of technology in search of a problem. Thus, scientists interested in sustainable agriculture who are involved and aware of basic biotechnological research may find themselves in a position to contribute a valuable problem-oriented focus to such research findings.

MARKER-ASSISTED BREEDING FOR
DURABLE PEST RESISTANCE IN INBRED CROPS

Many genes governing pest resistance are qualitative in nature and when widely deployed, impose tremendous selection pressure on pest populations. Compatible strains of the pest and pathogen populations may rapidly evolve to overcome the resistance. However, genes conferring qualitative resistance to pathogens that rapidly evolve resistance have been utilized for many years in conventional breeding programs because they could be most easily transferred into a variety of genetic backgrounds.

Molecular markers provide an opportunity to employ new strategies aimed at providing durable forms of resistance while utilizing traditional sources of resistance genes. The identification and utilization of markers closely linked to widely used resistance genes conferring qualitative and quantitative resistance makes it possible to efficiently move those genes into different genetic backgrounds. This is especially valuable when phenotypic evaluation for resistance is difficult or extremely tedious, that is, if the character is expressed late in the life of a plant (such as stalk rot in maize), occurs on an underground part of the plant (such as nematode resistance), or is especially difficult to measure (such as resistance to a quarantined pest).

In many cases, varieties with good adaptability and overall quality that have been accepted by growers as well as consumers will be appropriate targets for marker-assisted backcross breeding. Incremental improvements in pest resistance can be introduced in an efficient manner using markers (Tanksley et al., 1989). The ability to rapidly adjust existing varieties should allow breeders to respond more quickly to market demand.

One example of how molecular markers are influencing breeding strategies aimed at increasing yield stability and durability of resistance is the tagging of quantitative forms of resistance. Quantitative resistance, governed by several genes, is often more durable than resistance governed by single genes. Quantitative resistance may involve genes conferring both complete and partial resistance. Genes governing complete resistance inhibit pathogen reproduction on a host plant, whereas genes governing partial resistance reduce pathogen reproduction without completely suppressing it. Partial resistance imposes less selection pressure on the pathogen population in comparison to complete resistance and therefore reduces the probability of pathogen strains evolving that can overcome host resistance (Bonman et al., 1986; Chuke and Bonman, 1988). For this reason, breeders have had an interest in incorporating partial resistance into improved varieties, but phenotypic screening is difficult, both because of the quantitative nature of partial resistance and because it is less heritable than complete resistance.

In a recent study of durable blast resistance in a traditional West African variety of rice, Moroberekan, nine loci associated with both complete and partial resistance to the blast fungus *Pyricularia oryzae* were located via linkage to RFLP markers on the genetic map of rice (Wang et al., 1992). Some of the genes from this cultivar had been previously isolated by breeders, tagged with molecular markers, and were known to confer complete resistance to specific isolates of the fungus. In the traditional cultivar, genes linked to the same RFLP markers contributed to partial resistance. From this study we hypothesize that durable resistance in Moroberekan is the result of a combination of both complete and partial resistance factors and that these genes confer overlapping phenotypes, which make it virtually impossible to distinguish them phenotypically. With the aid of molecular markers, screening for the many genes controlling the resistance is technically feasible for the first time. New forms of polygenic resistance based on combinations of genes that affect pathogen reproduction in a variety of ways can now be developed.

Careful evaluation of pathogen population dynamics is a necessary prerequisite to developing effective pyramiding schemes or multilines, and each will have to be tailored to specific environments and pathogen populations. Markers capable of assessing pathogen diversity make it possible to monitor the effect of gene deployment on target pest populations (Gabriel et al., 1988; Hamer et al., 1989; Hartung and Civerolo, 1989; Lazo et al., 1987; Leach and White, 1991; Lee and Davis, 1988; Xu and Gonzalez, 1991). The objective of these efforts is to identify relationships among a variety of resistance genes in the host and virulence genes in pathogens, and between these genes and environmental factors that contribute to maintaining an acceptable balance between pest populations and crop production. Such a balance is the essence of durable resistance and an important component of sustainable agriculture.

In summary, biotechnological techniques can be used to increase the efficiency of breeding for resistance and facilitate the use of varied sources

173

and combinations of agronomically desirable genes, including genes from wild or unrelated species of organisms. By improving the durability of resistance, these techniques will also minimize the use of pesticides and fungicides.

HERBICIDES

"Of the 860 million pounds of pesticides used annually in the United States, estimates are that perhaps at most 1 percent reach their target pests. The rest simply contaminate soil, water, crops, and farm workers" (Goldburg, 1989). Herbicides comprise 60 percent of the pesticides used in the United States, or about 500 million pounds. Clearly a reduction in the use of herbicides or a shift to environmentally benign herbicides would greatly benefit the world environment.

Genetically engineered herbicide-resistant plants are currently a focus of research of some public universities as well as many private labs. Researchers have used molecular techniques to transfer genes conferring atrazine resistance from weeds into a variety of crops. Resistance to sulfonylurea compounds is conferred by introducing mutant acetolactate synthase genes into plants (Haughn et al., 1988). Glyphosate resistance has been incorporated into tomato and tobacco by Calgene, and petunia by Monsanto (Comstock, 1989).

What is the effect of herbicide-resistant crops on herbicide use? Proponents argue that herbicide-resistant crops will give farmers more weed-control options, lower overall herbicide use, and lower input costs (Goodman, 1989). They also point out that newer herbicides are much less toxic to humans and break down much more rapidly than the early versions of herbicides. Herbicide-resistant crops, however, may lead to increases in herbicide-resistant weeds through gene transfer and changes in patterns of herbicide use (Ellstrand and Hoffman, 1990).

New attitudes on the part of breeders, growers, and consumers are necessary to avoid perpetuating the old ideas of zero pest populations, 100% weed-free fields, and the importance of short-term gains in the face of longer-term losses. New forms of pest-resistant crop cultivars must be used with care to avoid the excessive dependence on a few highly effective genetic combinations that, if overused, will perpetuate the cycle of pest epidemics and dramatic yield loss. Combinations of different biotechnological approaches and conventional breeding schemes are contributing forms of pest-resistant crop plants that promise to be the building blocks of more sustainable agricultural systems. The effectiveness of these cultivars depends on how they are integrated into total production schemes.

Limitations of Biotechnology for Crop Improvement

Biotechnology offers considerable promise as an additional set of tools available to those breeding crop plants for sustainable systems. Molecular and cellular techniques can increase the speed and accuracy of selection, regenerate whole fertile plants from individual cells or organs, generate instant inbreds from cultured anthers, combine genes from unrelated organisms, and allow fine-tuned adjustments of gene expression. But there are limitations and concerns about the widespread use of these technologies.

A limitation to the immediate adoption of many biotechnological approaches is cost. In fact, the cost of the technology can be evaluated only relative to available alternatives. In cases of developing resistance to a quarantined pest, it may be more cost-effective to use molecular markers than initiate a conventional breeding program where the pest is endemic. Similarly, it may be impractical to transfer certain quantitatively inherited characters without the aid of molecular markers. In the area of seed health, molecular markers are very effective diagnostic tools for detecting minute levels of seed-transmitted viral, fungal, and bacterial pathogens. By providing a level of resolution never previously available, the cost of using these tools may be justified.

Although it is true that many cellular and molecular techniques are very expensive, it is also true that the technology is becoming more affordable and technically simpler. The advent of PCR-based forms of molecular markers is a dramatic simplification of its forerunner, RFLP technology. The initial identification, cloning, and vector construction required to genetically engineer a crop plant represents a large initial budgetary and technical commitment but holds the potential for broad application. A gene cloned from the cucumber mosaic virus (CMV) is currently being used to transform cucumber, melon, squash, and pepper, providing CMV resistance to these crops.

A different level of concern over the use of biotechnology has to do with the fear of tinkering with nature. This fear is not unique to the age of biotechnology. New technology has always inspired distrust and in many cases this distrust imposes a healthy set of checks and balances on the scientific community.

Several potentially deleterious environmental impacts have been postulated in relation to widespread acceptance of genetic engineering in crop improvement. These include the inadvertent transfer of alien genes to a wild relative through natural outcrossing; the displacement of natural communities of microbes, insects, or plants; the accidental escape of an untested engineered organism; and overdependence on simply inherited engineered resistance characters (Hoffman, 1990; Klingmuller, 1988). Rational concerns must be evaluated taking proper precautions, so that over time the validity of these concerns can be determined (Tiedje et al., 1989). In fact, precautions and guidelines have been established as a result of broad skepticism in the public and scientific communities regarding the use of genetic engineering in agriculture (Persley and Peacock, 1990). The U.S. Department of Agriculture, Food and Drug Administration, and Environmental Protection Agency are developing a coordinated framework for regulating genetically engineered crops and food products (Fraley, 1992). In Europe, national regulations still vary, despite a European Economic Community–wide framework policy on field release.

Developing checks and balances in all of our crop research and food production efforts is of critical importance. Familiarity with the technology, rational evaluation of research goals, clear oversight of labs working with genetically engineered organisms, and careful containment of genetically engineered organisms that represent a health or environmental hazard, are all important checks on the research community and ensure that biotechnology research will be conducted in a responsible fashion.

A different form of ecological disruption may be envisioned where biotechnology successfully contributes to generating crop cultivars that are productive in environments previously considered unfit for agriculture. In this case, marginal and ecologically vulnerable lands are endangered through cultivation of improved disease- or stress-tolerant cultivars.

Areas such as mangrove swamps and tidal wetlands are fragile and complex ecosystems that cannot support intensive cultivation as we know it. With the assistance of biotechnology, rice varieties with resistance to salt, acidity, and submergence may be developed that yield well in these areas, jeopardizing the ecosystem. In the face of human demand for food, the fragility and importance of these habitats may be ignored as crop production is pushed to its limits. The result of increased production in these zones, it is argued, would be to encourage human populations to overload the carrying capacity of the ecosystem, totally disrupting the intricate web of organisms, nutrients, and habitats that make these environments what they are and diminishing the quality of life for those who inhabit the region.

As population pressure in many countries forces thousands or millions of people to marginal lands, the social and political components of food production become as important as the biological ones. Though biotechnology can be useful in developing crop cultivars suited to marginal areas, it is clearly not a technological issue that determines whether or not populations are able to focus on long-term objectives such as the sustainability of their agriculture or whether they are caught in the treadmill of "food at any cost." Policy decisions relating to population control and land management rather than biotechnology and varietal improvement hold the key to these issues of sustainability.

In its broadest perspective, sustainable agriculture must take into account the welfare of both the producers and consumers of agricultural commodities. Sometimes the interests of these two groups are diametrically opposed. Consumers want inexpensive food and fiber whereas farmers involved in a cash economy benefit from high prices. Sustainable agriculture depends on some measure of ecological, economic, and social predictability. Plant breeders work within all three spheres and set objectives appropriate to the time and place in which they work and live. One thing they as a group must unfailingly do is to encourage diversity. No single approach or method is likely to be totally successful, and no single cultivar can meet all the world's needs.

Amid the diversity of plant breeding strategies and objectives emerges the question of how the advent of biotechnology in agriculture will influence the relationship between the private sector, producers, and consumers. Genetically engineered seed is likely to decrease the investment in agricultural inputs such as pesticides, but how much will the farmer have to pay to buy this seed? Will the cost and productivity of genetically engineered seed simply put marginally productive farmers out of business, because the capital required to buy the seed becomes a limiting factor and production soars as a result of breeding advances? Will farmers who cannot obtain or afford such improved seed be able to continue their old ways of life?

How broadly will these biotechnologies be applied for commodities that are not highly profitable for seed companies? In order to justify the cost of an investment in an improvement, a company typically hopes to develop a characteristic that is exclusively theirs in order to sell the improved seed at a price that recoups research costs. How many really useful improvements are also patentable? Will the most useful applications of biotechnology remain unavailable to the farmer because a company cannot be assured of exclusive rights to the product, and therefore the profit?

There is a clear need for better links to be established between molecular biologists in both the public and the private sector and those who focus on the long-term objectives of ecological, nutritional, and economic sustainability in agricultural systems. Biotechnology is good at offering specific and often innovative solutions to well-defined problems. Defining problems that clearly impact on the issues of sustainability, equity, and profitability of food production, and justify the time, effort, and expense involved in generating biotechnological solutions is one of the most formidable challenges facing plant breeders and molecular biologists alike.

Historically, communication between field-oriented plant breeders and lab-oriented cell and molecular biologists has been poor. This situation is changing as the techniques of biotechnology become more routine and promises of immediate productivity and profitability are deflated. The bottom line is that extensive field testing of new cultivars is required to convincingly test the results of any development aimed at crop improvement, regardless of the source of that development.

Genetic manipulation generally requires the skills of an experienced molecular or cell biologist, but field testing over years and environments is typically the domain of the conventional plant breeder and is the only avenue for getting a product into the hands of a farmer. The integration of biotechnology and traditional plant improvement methods can be greatly facilitated by the exchange of technology and human resources.

In an international context, the international crop improvement centers may be instrumental in coordinating such efforts. An example of an effective cooperative approach is provided by the Rockefeller Foundation's program on rice biotechnology. The Foundation provided support to a group of the world's leading plant molecular biology and cell biology laboratories with the objective of stimulating biotechnology research on rice to achieve a level comparable with biotechnological research on the major cereals of the industrial world (Persley and Peacock, 1990). As the technology has been developed, the emphasis of the program has shifted to the transfer of biotechnology to developing country laboratories and applied rice improvement programs. The ultimate success of this effort will be judged by the impact of biotechnology on increasing and sustaining rice production in developing countries.

References

Athwal, D.S., M.D. Pathak, E.H. Bacalangco, and C.D. Pura. 1971. Genetics of resistance to brown planthoppers and green leafhoppers in *Oryza sativa* L. Crop Sci. 11:747–750.

Barbier, P., H. Morishima, and A. Ishihama. 1991. Phylogenetic relationships of annual and perennial wild rice: Probing by direct DNA sequencing. Theor. Appl. Genet. 81:693–702.

Beachy, R.N., T. Qu, P. Shen, A. deKochko, M. Bhattachuryya, G. Laco, D. Chen, M. Kaniewska, and C. Fauquet. 1991. Rice tungro disease: Genome organization of RTSV and RTBV and isolation of the CP genes.

p.88. *In* Fifth Annu. Meeting Int. Program Rice Biotechnol. Abstracts. Rockefeller Foundation, New York, NY.

Beachy, R.N., S. Loesch-Freios, and N.G. Turner. 1990. Coat protein-mediated resistance against virus infection. Annu. Rev. Phytopathology 28:451–474.

Bernatzky, R., and S.D. Tanksley. 1986. Toward a saturated linkage map in tomato based on isozymes and random dDNA sequences. Genetics 112:887–898.

Bhojwani, S.S., and M.K. Razdan. 1983. Plant tissue culture: Theory and practice. Biodiversity Conservation Strategy Update. 1991. 2(1):1–10.

Biodiversity Conservation Strategy Update. Vol. 2. No. 1, Summer 1991. World Resources Institute, Washington, DC.

Bonierbale, M.W., R.L. Plaisted, and S.D. Tanksley. 1988. RFLP maps based on a common set of clones reveal modes of chromosomal evolution in potato and tomato. Genetics 120:1093–1103.

Bonman, J.M., B.A. Estrada, and R.I. Denton. 1986. Blast management with upland rice cultivar mixtures. p.375–382. *In* Progress in upland rice research. Proc. of conference held in Jakarta, Indonesia. Feb. 1985.

Botstein, D., R.L. White, M. Skolnick, and R.W. Davis. 1980. Construction of a genetic linkage map in man using restriction fragment length polymorphisms. Am. J. Hum. Genet. 32:314–331.

Brar, D.S., R. Elloran, and G.S. Khush, 1991. Interspecific hybrids produced through embryo rescue between cultivated and eight wild species of rice. Rice Genet. Newsl. 8:91–93.

Buckner, G.D., and J.K. Kastle. 1917. The growth of isolated plant embryos. J. Biol. Chem. 29:209–213.

Cao, J., W. Zhang, D. McElroy, and R. Wu. 1991. Assessment of rice genetic transformation techniques. p.175–198. *In* G.S. Khush and G.H. Toenniessen (ed.) Rice Biotechnology. Biotechnology in Agriculture Series No. 6. Commonwealth Agricultural Bureaux, Wallingford, Oxon, U.K.

Chang, T.T. 1984. Conservation of rice genetic resources: Luxury or necessity? Science 224:251–256.

Chuke, K.C., and J.M. Bonman. 1988. Changes in virulence frequencies of *Pyricularia oryzae* in pure and mixed stands of rice. J. Plant Protection Tropics 5(1):23–29.

Collmer, C.W., and S.H. Howell. 1992. Role of satellite RNA in the expression of symptoms caused by plant viruses. Annu. Rev. Phytopathology 30:419–442.

Comstock, G. 1989. Is genetically engineered herbicide-resistance (GEHR) compatible with low-input sustainable agriculture (LISA)? p.111–123. *In* J. Fessenden-MacDonald (ed.) NABC Report 1: Biotechnology and sustainable agriculture: Policy alternatives. Boyce Thompson Institute, Ithaca, NY.

Cotten, M. 1990. The *in vivo* application of ribozymes. Trends in Biotechnol. 8(7):174–178.

Dally, A., and G. Second. 1990. Chloroplast DNA diversity in wild and cultivated species of rice (Genus *Oryza,* Section *Oryza*). Cladistic-mutation and genetic-distance analysis. Theor. Appl. Genet. 80:209–222.

Deroles, S.C., and R.C. Gardner. 1988. Expression and inheritance of kanamycin resistance in a large number of transgenic petunias generated by *Agrobacterium*-mediated transformation. Plant Mol. Biol. 11:355–364.

de Vicente, M.C. 1991. Genetic studies in a wide cross of tomato using RFLPs: Heterosis, transgression, and sex-recombination. Ph.D. diss. Cornell University, Ithaca, NY.

Dilday, R.H. 1990. Contribution of ancestral lines in the development of new cultivars of rice. Crop Sci. 30:905–911.

Dixon, R.A., and C.J. Lamb. 1990. Molecular communication in plant: microbial pathogen interactions. Annu. Rev. Plant Physiol. 41:339–367.

Ellstrand, N.C., and C.A. Hoffman. 1990. Hybridization as an avenue of escape for engineered genes. Bioscience 40:438–442.

Endo, T., and H. Morishma. 1983. Rice. p.129–146. *In* S.D. Tanksley and T.J. Orton (ed.) Isozymes in plant genetics and breeding. Part B. Elsevier, Amsterdam.

Evola, S.V., F.A. Burr, and B. Burr. 1986. The suitability of restriction fragment length polymorphisms as genetic markers in maize. Theor. Appl. Genet. 71:765–771.

Fatokun, C.A., D.I. Menancio-Hautea, D. Danesh, and N.Y. Young, 1992. Evidence for othologous seed weight genes in cowpea and mung bean based on RFLP mapping. Genetics 132:841–846.

Fraley, R. 1992. Sustaining the food supply. Biotechnol. 10:40–43.

Gabriel, D.W., J.E. Hunter, M.T. Kingsley, J.W. Miller, and G.R. Lazo. 1988. Clonal population structure of *Xanthomonas campestris* and genetic diversity among citrus canker strains. Mol. Plant-Microbe Interactions 1(2):59–65.

Ganal, M.W., G.B. Martin, R. Messeguer, and S.D. Tanksley. 1990. Application of RFLPs, physical mapping and large DNA technologies to the cloning of important genes from crop plants. AgBiotech News Information 2:835–840.

Gasser, C.S., and R.T. Fraley. 1990. Genetically engineering plants for crop improvement. Science 244:1293–1299.

Glaszmann, J.C. 1987. Isozymes and classification of Asian rice varieties. Theor. Appl. Genet. 74:21–30.

Godshalk, E.B., M. Lee, and K.R. Lamkey. 1990. Relationship of restriction fragment length polymorphisms to single-cross hybrid performance of maize. Theor. Appl. Genet. 80:273–280.

Goldburg, R.J. 1989. Should the development of herbicide-tolerant plants be a focus of sustainable agriculture research? p.103–110. *In* J. Fessenden-MacDonald (ed.). NABC Report 1: Biotechnology and sustainable agriculture: Policy alternatives. Boyce Thompson Institute, Ithaca, NY.

Goldberg, L.J., and J. Margalif. 1977. A bacterial spore demonstrating rapid larvicidal activity against *Anopheles serengetii, Uranotaenia unguiculata, Culex univittatus, Aedes aegypti,* and *Culex pipiens.* Mosquito News 37:355–358.

Golemboski, D.B., G.P. Lomonossoff, and M. Zaitlin. 1990. Plants transformed with a tobacco mosaic virus nonstructural gene sequence are resistant to the virus. Proc. Natl. Acad. Sci. USA 87:6311–6315.

Goodman, R.M. 1989. Policy alternatives in sustainable agriculture. p.48–57. *In* J. Fessenden-MacDonald (ed.). NABC Report 1. Biotechnology and sustainable agriculture: Policy Alternatives. Boyce Thompson Institute, Ithaca, NY.

Gordon-Kamm, W.J., T.M. Spencer, M.L. Mangano, T.R. Adams, R.J. Daines, W.G. Start, J.V. O'Brien, S.A. Chambers, W.R. Adams, Jr., N.G. Willetts, T.B. Rice, C.J. Mackey, R.W. Kureger, A.P. Kausch, and P.G. Lemaux. 1990. Transformation of maize cells and regeneration of fertile transgenic plants. The Plant Cell 2:603–618.

Gould, F. 1988. Evolutionary biology and genetically engineered crops. Bioscience 38:26–33.

Grosser, J.W., and F.G. Gmitter, Jr. 1991. Protoplast technology in tropical fruit improvement, with focus on *Citrus*. *In* Proc. Workshop Agric. Biotechnol. for Enhancing Agric. Development in Indonesia. Bogor, Indonesia. 21–24 May 1991. Agency for Agricultural Research and Development, Bogor, Indonesia.

Guiderdoni, E., J.C. Glaszmann, and B. Courtois. 1989. Segregation of 12 isozyme genes among doubled haploid lines derived from a japonica x indica cross of rice (*Oryza sativa* L.). Euphytica 42:45–53.

Hamer, J.E., L. Farrall, M.J. Orbach, B. Valent, and F.G. Chumley. 1989. Host species-specific conservation of a family of repeated DNA sequences in the genome of a fungal plant pathogen. Proc. Natl. Acad. Sci. USA 86:9981–9985.

Hargrove, T.R., V.L. Cabanilla, and W.R. Coffman. 1988. Twenty years of rice breeding. Bioscience 38:675–681.

Hartung, J.S., and E.L. Civerolo. 1989. Restriction fragment length polymorphisms distinguish *Xanthomonas campestris* strains isolated from Florida citrus nurseries from *X.C.* pv. *citri*. Mol. Plant Pathology 79(7): 793–799.

Haseloff, J., and W.L. Gerlach. 1988. Simple RNA enzymes with new and highly specific endoribonuclease activities. Nature 334:585–591.

Haughn, G.W., J. Smith, B. Mazur, and C. Somerville. 1988. Transformation with a mutant *Arabidopsis* acetolactate synthase gene renders tobacco resistant to sulfonylurea herbicides. Mol. General Genet. 211: 266–271.

Helentjaris, T., G. King, M. Slocum, C. Siedenstrang, and S. Wegman. 1985. Restriction fragment polymoprhisms as probes for plant diversity and their development as tools for applied breeding. Plant Mol. Biol. 5:109–118.

Helentjaris, T. 1987. A genetic linkage map for maize based on RFLPs. Trends Genet. 3:217–221.

Heun, M., A.E. Kennedy, J.A. Anderson, N.L.V. Lapitan, M.E. Sorrells, and S.D. Tanksley. 1991. Construction of an RFLP map for barley (*Hordeum vulgare*). Genome 34:437–447.

Hilder, V.A., A.R.R. Gatehouse, S.E. Sheerman, R.F. Barker, and D. Boulter. 1987. A novel mechanism of insect resistance engineered into tobacco. Nature 330:160–163.

Hodges, T.K., J. Peng, L.A. Leszek, and D.S. Koetje. 1991a. Transformation and regeneration of rice protoplasts. p.157–174. *In* G.S. Khush and G.H. Toenniessen (ed.) Rice biotechnology. Biotechnology in agriculture No. 6. Commonwealth Agricultural Bureaux, Wallingford, Oxon, U.K.

Hodges, T.K., J.Y. Peng, R.C. Su, F. Wen, H. Kononowicz, D. Koetje, X.Q. Li, and M. Rudert. 1991b. Indica rice regeneration and transformation. p.47. *In* Fifth Annu. Meeting Int. Program Rice Biotechnol. Abstracts. Rockefeller Foundation, New York, NY.

Hoffman, C.A. 1990. Ecological risks of genetic engineering of crop plants. Bioscience 40(6):434–437.

Hofte, H., J. Seurnick, A. Van Houtven, and M. Vaeck. 1987. Nucleotide sequence of a gene encoding an insecticidal protein of *Bacillus thuringiensis* var. *tenebrionis* toxic against Coleoptera. Nucleic Acids Res. 15:7183–7186.

Hulbert, S.H., T.W. Ilott, E.J. Legg, S.E. Lincoln, E.S. Lander, and R.W. Michelmore. 1988. Genetic analysis of the fungus *Bremia lactucae*, using restriction fragment length polymorphisms. Genetics 120:947–958.

Hull, R., J. Hay, I. Dasgupta, M. Blakebrough, G. Lee, J.W. Davies, G. Dahal, Y. Bao, and Z. Fan. 1991. The molecular biology of rice tungro bacilliform virus. p.89. *In* Fifth Annu. Meeting Int. Program Rice Biotechnol. Abstracts. Rockefeller Foundation, New York, NY.

Ishii, T., T. Terachi, and K. Tsunewaki. 1986. Restriction endonuclease analysis of chloroplast DNA from cultivated rice species *Oryza sativa* and *Oryza glaberrima*. Jpn. J. Genet. 61(6):537–541.

Ishii, T., T. Terachi, and K. Tsunewaki. 1988. Restriction endonuclease analysis of chloroplast DNA from A-genome diploid species of rice. Jpn. J. Genet. 63(6):523–536.

Ivanovskaya, E.V. 1946. Hybrid embryos of cereals grown on artificial nutrient medium. C.R. (Dokl.) Acad. Sci. URSS 54:445–448.

Iyer, R.D., and O.P. Govila. 1964. Embryo culture of interspecific hybrids in the genus *Oryza*. Indian J. Genet. Plant Breed. 24:116–121.

Izant, J.G. 1989. Antisense "Pseudogenetics": Cell motility and the cytoskeleton 14:81–91.

Johal, G., and S.P. Briggs. 1991. Molecular cloning of the Hm1 disease resistance gene in maize. Abstract 1384. *In* R.B. Hallick (ed.) Third Int. Congr. Plant Mol. Biol. Tucson, AZ. 6–11 Oct., 1991. Univ. Arizona, Tucson, AZ.

Kam-Morgan, L.N.W., B.S. Gill, and S. Muthukrishnan. 1989. DNA restriction fragment length polymorphisms: A strategy for genetic mapping of D genome of wheat. Genome 32:724–732.

Kaniewski, W.K., C. Lawson, B. Sammons, L. Haley, J. Hart, X. Delannay, and N.E. Tumer. 1990. Field resistance of transgenic Russet Burbank potato to effects of infection by potato virus X and potato virus Y. Biotechnol. 8:750–754.

Klingmuller, W. (ed.) 1988. Risk assessment for deliberate releases: The possible impact of genetically engineered microorganisms on the environment. Springer-Verlag, Berlin and Heidelberg, Germany.

Khush, G.S. 1984. Breeding rice for resistance to insects. Protection Ecology 7:147–165.

Lambert, B., and M. Peferoen. 1992. Insecticidal promise of *Bacillus thuringiensis*. Bioscience 42(2):112–122.

Landry, B.S., V. Rick, B.F. Kesseli, and R.W. Michelmore. 1987. A genetic map of lettuce (*Lactuca sativa* L.) with restriction fragment length polymorphism, isozyme, disease resistance, and morphological markers. Genetics 116:331–337.

Lazo, G.R., R. Roffey, and D.W. Gabriel. 1987. Pathovars of *Xanthomonas campestris* are distinguishable by restriction fragment-length polymorphism. Int. J. Systematic Bacteriol. 37:214–221.

Leach, J.E., and F.F. White. 1991. Molecular probes for disease diagnosis and monitoring. p.281–308. *In* G.S. Khush and G.H. Toenniessen (ed.) Rice biotechnology. Biotechnology in Agriculture Series No. 6. Commonwealth Agricultural Bureaux, Wallingford, Oxon, U.K.

Lee, I.M., and R.E. Davis. 1988. Detection and investigation of genetic relatedness among aster yellows and other mycoplasmalike organisms by using cloned DNA and RNA probes. Mol. Plant-Microbe Interactions 1(8):303–310.

Leung, H., and M. Taga. 1988. *Magnaporthe grisea* (*Pyricularia* species), the blast fungus. Adv. Plant Pathology 6:175–188.

Li, Z., S. Pinson, W.D. Park, and J.W. Stansel. 1991. Mapping QTLs for yield, quality, and disease resistance in rice using RFLP markers. p.130. *In* Fifth Annu. Meeting Int. Program Rice Biotechnol. Abstracts. Rockefeller Foundation, New York, NY.

Liu, Y.G., N. Mori, and K. Tsunewaki. 1990. Restriction fragment length polymorphism (RFLP) analysis in wheat. 1. Genomic DNA library construction and RFLP analysis in common wheat. Jpn. J. Genet. 65(5): 367–380.

Martin, G., J.G.K. Williams, and S.D. Tanksley. 1990. Rapid identification of markers linked to a *Pseudomonas* resistance gene in tomato by using random primers and nearly isogenic lines. Proc. Natl. Acad. Sci. USA 88:2336–2340.

Matzke, M.A., M. Primig, J. Trnovsky, and A.J.M. Matzke. 1989. Reversible methylation and inactivation of marker genes in sequentially transformed tobacco plants. EMBO J. 8:643–649.

McCouch, S.R., J.L. Abenes, R. Angeles, G.S. Khush, and S.D. Tanksley. 1991. Molecular tagging of a recessive gene, xa-5, for resistance to bacterial blight of rice. Rice Genet. Newsl. 8: 143–145.

McCouch, S.R., G. Kochert, Z.H. Yu, Z.Y. Wang, G.S. Khush, W.R. Coffman, and S.D. Tanksley. 1988. Molecular mapping of rice chromosomes. Theor. Appl. Genet. 76:815–829.

McPherson, S.A., F.J. Perlak, R.L. Fuchs, P.G. Marrone, P.B. Lavrik, and D.A. Fischhoff. 1988. Characterization of the coleopteran-specific protein gene of *Bacillus thuringiensis* var. *tenebrionis*. Biotechnol. 6:61–66.

Meadows, J., S.S. Gill, and L.W. Bone. 1990. *Bacillus thuringiensis* strains affect population growth of the free-living nematode *Turbatrix aceti*. Invertebr. Reprod. Dev. 17:73–76.

Miller, J.C., and S.D. Tanksley. 1990. RFLP analysis of phylogenetic relationships and genetic variation in the genus *Lycopersison*. Theor. Appl. Genet. 80:437–448.

Murray, M.G., Y. Ma, J. Romero-Severson, D.P. West, and J.H. Cramer. 1988. Restriction fragment length polymorphisms: What are they and how can breeders use them? Proc. Annu. Corn Sorghum Ind. Res. Conf. 43:72–87.

National Academy of Sciences. 1972. Genetic vulnerability of food crops. National Academy Press, Washington, DC.

Nelson, R.S., S.M. McCormick, X. Delannay, P. Dube, J. Layton, E. Anderson, M. Kaniewski, R.K. Proksh, R.B. Horsch, S.G. Rogers, R.T. Fraley, and R.N. Beachy. 1988. Virus tolerance, plant growth, and field performance of transgenic tomato plants expressing coat protein from Tobacco Mosaic Virus. Biotechnol. 6:403–409.

Nishizawa, Y. and T. Hibi. 1991. Rice chitinase gene: cDNA cloning and stress induced expression. Plant Sci. 76:211–218.

Ottaviano, E., M.S. Gorla, E. Pe, and C. Frova. 1991. Molecular markers (RFLPs and HSPs) for the genetic dissection of thermotolerance in maize. Theor. Appl. Genet. 81:713–719.

Paszkowski, J., R.D. Shillito, M. Saul, V. Mandak, T. Hohn, B. Hohn, and I. Potrykus. 1984. Direct gene transfer to plants. EMBO J. 3(12):2717–2722.

Paterson, A.H., S. Damon, J.P. Hewitt, D. Zamir, H.D. Rabinowitch, S.E.

Lincoln, E.S. Lander, and S.D. Tanksley. 1991. Mendelian factors underlying quantitative traits in tomato: Comparison across species, generations, and environments. Genetics 127:181–197.

Paterson, A.H., E.S. Lander, J.D. Hewitt, S. Peterson, S.E. Lincoln, and S.D. Tanksley. 1988. Resolution of quantitative traits into Mendelian factors by using a complete linkage map of restriction fragment length polymorphisms. Nature 335:721–726.

Perlak, F.J., R.L. Fuchs, D.A. Dean, S.L. McPherson, and D.A. Fischhoff. 1991. Modification of the coding sequence enhances plant expression of insect control protein genes. Proc. Natl. Acad. Sci. USA 88: 3324–3328.

Persley, G.J., and W.J. Peacock. 1990. Biotechnology for bankers. p.3–24. *In* G.J. Persley (ed.) Agricultural biotechnology: Opportunities for international development. Commonwealth Agricultural Bureaux, Wallingford, Oxon, U.K.

Plucknett, D.L., J.I. Cohen, and M.E. Horne. 1990. Role of international agricultural research centres. p.400–414. *In* G.J. Persley (ed.) Agricultural biotechnology: Opportunities for international development. Commonwealth Agricultural Bureaux, Wallingford, Oxon, U.K.

Powell-Abel, P.A., R.S. Nelson, T. De, N. Hoffman, S.G. Rogers, R.T. Fraley, and R.N. Beachy. 1986. Delay of disease development in transgenic plants that express the tobacco mosaic virus coat protein gene. Science 232:738–743.

Ronald, P.C., B. Albano, R. Tabien, L. Abenes, K.S. Wu, S. McCouch, and S.D. Tanksley. 1992. Genetic and physical analysis of the rice bacterial blight disease resistance locus, *Xa21*. PNAS (in press).

Rossi, J.J., and N. Sarver. 1990. RNA enzymes (ribozymes) as antiviral therapeutic agents. Trends Biotechnol. 179:183.

Second, G.A. 1985. Evolutionary relationships in the sativa group of *Oryza* based on isozyme data. Genet. Sel. Evol. 17(1):89–114.

Second, G.A. 1982. Origin of the genic diversity of cultivated rice (*Oryza* spp.): Study of the polymorphism scored at 40 isozyme loci. Jpn. J. Genet. 57:25–57.

Shattuck-Eidens, D.M., R.N. Bell, S.L. Neuhausen, and T. Helentjaris. 1990. DNA sequence variation within maize and melon: Observations from polymerase chain reaction amplification and direct sequencing. Genetics 126:207–217.

Shimamoto, K., R. Terada, J. Izawa, and H. Fujimoto. 1989. Fertile transgenic rice plants regenerated from transformed protoplasts. Nature 338:274–276.

Sitch, L.A., R.D. Dalmacio, and G.O. Romero. 1989. Crossability of wild *Oryza* species and their potential use for improvement of cultivated rice. Rice Genet. Newsl. 6:58–60.

Stuber, C.W., S.E. Lincoln, D.W. Wolff, T. Helentjaris, and E.S. Lander. 1992. Identification of genetic factors contributing to heterosis in a hybrid from two elite maize inbred lines using molecular markers. Genetics 132:823–839.

Tabashnik, B.E., N.L. Cusdhing, N. Finson, and M.W. Johnson. 1990. Field development of resistance to *Bacillus thuringiensis* in Diamondback Moth (Lepidoptera: Plutellidae). J. Econ. Entomol. 83(5): 1671–1676.

Tanksley, S.D., R. Bernatzky, N.L. Lapitan, and J.P. Prince. 1988. Conservation of gene repertoire but not gene order in pepper and tomato. Proc. Natl. Acad. Sci. USA 85:6419–6423.

Tanksley, S.D., N.D. Young, A.H. Paterson, and M.W. Bonierbale. 1989. RFLP mapping in plant breeding: New tools for an old science. Biotechnol. 7:257–264.

Tiedje, J.M, R. K. Colwell, V.L. Grossman, R.E. Hodson, R.E. Lenski, R.N. Mack, and P.J. Regal. 1989. The planned introduction of genetically engineered organisms: Ecological considerations and recommendations. Ecology 70:298–315.

Tien, P., and G. Wu. 1991. Satellite RNA for the biocontrol of plant disease. Adv. Virus Res. 39:321–339.

Tingey, S., S. Sebastian, and J.A. Rafalski. 1989. An RFLP map of the soybean genome. p.180. *In* Agronomy abstracts. Am. Soc. Agron., Madison, WI.

Toenniessen, G.H. 1991. Potentially useful genes for rice genetic engineering. p.253–280. *In* G.S. Khush and G.H. Toenniessen (ed.) Rice biotechnology. Biotechnology in Agriculture Series No. 6. Commonwealth Agricultural Bureaux, Wallingford, Oxon, U.K.

Uchimiya, H., T. Handa, and D.S. Brar. 1989. Transgenic plants. J. Biotechnol. 12:1–20.

Walbot V., and D. Gallie. 1991. Gene expression in rice. p.225–252. *In* G.S. Khush and G.H. Toenniessen (ed.) Rice biotechnology. Biotechnology in Agriculture Series No. 6. Commonwealth Agricultural Bureaux, Wallingford, Oxon, U.K.

Wang G. , D.J. Mackill, J.M. Bonman, S.R. McCouch, and R.J. Nelson. 1992. RFLP mapping of qualitative and quantitative genes for blast resistance in a durably resistant rice cultivar. Genetics (accepted).

Wang, Z.Y., and S.D. Tanksley. 1989. Restriction fragment length polymorphism in *Oryza sativa* L. Genome 321:1113–1118.

Wang, Z.Y., G.A. Second, and S.D. Tanksley. 1992. Polymorphism and phylogenetic relationships among species in the genus *Oryza* as determined by analysis of nuclear RFLPs. Theor. Appl. Genet. 82 (in press).

Whitten, M.J., and J.G. Oakeshott. 1990. Biocontrol of insects and weeds. p.123–142. *In* G.J. Persley (ed.) Agricultural biotechnology: Opportunities for international development. Commonwealth Agricultural Bureaux, Wallingford, Oxon, U.K.

Williams, J.G.K., A.R. Kubelik, K.J. Livak, J.A. Rafalski, and S.V. Tingey. 1991. DNA polymorphisms amplified by arbitrary primers are useful genetic markers. Nucleic Acids Res. 18:6531–6535.

Xu, G.W., and C.F. Gonzalez. 1991. Plasmid, genomic, and bacteriocin diversity in U.S. strains of *Xanthomonas campestris* pv. *oryzae*. Phytopathology 81:628–631.

Yeung, E.C., T.A. Thorpe, and C.J. Jensen. 1981. In vitro fertilization and embryo culture. p.253–271. *In* T.A. Thorpe (ed.) Plant tissue culture: Methods and applications in agriculture. Academic Press, New York, NY.

Yu, Z.H., D.J. Mackill, J.M. Bonman, and S.D. Tanksley. 1991. Tagging genes for blast resistance in rice via linkage to RFLP markers. Theor. Appl. Genet. 81:471–476.

Zhu, Q., and C.J. Lamb. 1991. Isolation and characterization of a rice gene encoding a basic chitinase. Mol. Gen. Genet. 226:289–296.

10

Genotype by Environment Interaction
in Crop Improvement

Ed Souza, Jim R. Myers, Brian T. Scully

The phenotype of a plant is the observed expression of the genotype in response to the environment. Cultivars of differing genetic composition may respond differently when placed in various growing conditions. The differential response of genotypes to the environment is termed *Genotype by Environment Interaction* (GxE). GxE limits the value of selections made among breeding lines within a single environment. The phenotypic performance of a breeding line in a single environment does not necessarily provide an accurate assessment of the genetic potential of the line. Thus GxE reduces a breeder's ability to correctly choose a superior genotype. The testing program of a crop improvement project must determine the environmental response of a genotype to validate performance before release to the public.

Sustainable agriculture is defined for this chapter as a cropping system with a management regime, including varietal selection, which minimizes purchased inputs from outside the farm and maximizes long-term economic return with minimum impact to the environment. Breeding for sustainable agriculture poses unique GxE problems. Within a sustainable cropping system the performance of a genotype may need to be measured under a

This chapter is Paper No. 91791, Dep. Plant, Soil, and Entomological Sci., Univ. Idaho, Moscow, ID, and Florida Agric. Expt. Sta. Journal Series No. R-02150.

broad range of soil fertilities, cropping rotations, biotic stresses, or species mixtures. In a high-input cropping system external inputs are typically applied to buffer biotic and abiotic stresses; the inability of a genotype to tolerate stress may not lead to an unreliable cultivar if stress is managed agronomically.

A significant body of sustainable agriculture research is directed towards developing countries. Inasmuch as many developing countries are located in the tropics or subtropics, they often have more diverse environments than those found in temperate zones. The diversity of environments reflects a tremendous climatic variation over very short distances and provides an opportunity for crop cultivation in different seasons of the year. Growers may use the same cultivars for all production seasons and thus require adaptation to all seasons (Unander et al., 1989). In addition, many growers in the tropics are at a level of subsistence where crop failure is disastrous. Such growers would choose cultivars with the lowest associated risk.

Within a testing program, environments should be as diverse and adverse as the conditions of the target environments. Stability of plant phenotype based on trials that are unrepresentative of target environments may lead to an inaccurate assessment of a cultivar's vulnerabilities. Therefore, phenotypic stability estimates for grower use should be derived from on-farm testing whenever possible (Hildebrand, 1984). Measurement of the genotype is well understood. Measurement of the environment in a manner descriptive of plant responses is difficult. This difficulty will likely increase as cropping systems become more sustainable, which implies increased ecological complexity. Therefore, the assessment of the environmental component of the G x E must also advance.

This chapter focuses on (1) typical interactions of genotypes with the environment, (2) description of models used for measuring G x E, (3) integration of those models into a breeding program, and (4) selection methodologies for breeding phenotypically stable cultivars for sustainable systems. The G x E literature has been reviewed by Lin et al. (1986), Becker and Leon (1988), Kang (1990), and Romagosa and Fox (in press). These publications provide a broad range of information on models used for measuring and interpreting GxE. Few of the models have been discussed specifically for their use in sustainable agriculture (Smith et al., 1990). The intent of this chapter is to apply the most typical models and GxE interpretations to the development of crops for sustainable agriculture.

Types of Interactions between Genotype and Environment

Most cultivar development programs test at several locations per year. Locations are often assumed to be equivalent to years for the measurement of GxE. It is likely that different components of GxE will be emphasized by the combination of years or locations that are used in the analysis. Thus it is important to determine if random or fixed effects are more important in their influence on GxE. Rasmussen and Lambert (1961) found genotype by year interaction to be more important than genotype by location interaction for barley. On the other hand, Kang and Gorman (1989) found that site was a more important contributor to maize GxE than was year. These findings are representative of the general lack of agreement on the importance of locations vs. years for GxE. (see Jensen, 1988, p.608–614). Jensen's summary indicates that the optimum mix of locations and years is unique and the proper analysis must be arrived at empirically. This may be doubly true for sustainable agricultural systems where the body of breeding literature for a species may not be particularly applicable to new cropping patterns.

Stability can be defined as the reliable performance of a cultivar over a range of environments (Becker and Leon, 1988). An *environment* is a crop site within a location and year and is considered to be a unique growing condition, representative of the target growing conditions of farmers' fields. In practice, stability may be conceptually different depending upon the statistical method employed. A review and subsequent paper by Lin and co-workers defined four types of stability (Lin et al., 1986; Lin and Binns, 1988a). These are biological, or static, stability (Type 1); agronomic, or dynamic, stability (Type 2); linear stability (Type 3); and temporal stability (Type 4).

A genotype has Type 1 stability if genotypic variance across environments is small (Lin et al., 1986). Genotypic variances and coefficients of

194

variation have been used to describe this type of stability. Linear regression (the mean of a genotype predicted by the mean of all genotypes at a location, Finlay and Wilkinson, 1963) has been commonly used to define Type 1 stability where perfect stability would be a slope of 0. In the context of sustainable agriculture, biological stability is certainly attractive because cultivars bred for low-input situations will need to perform well in the face of stress not experienced by cultivars in conventional agriculture. Biologically stable, or homeostatic, cultivars do not respond favorably to improved cropping systems or environments but have the capacity to buffer adverse environments (Baker, 1981).

Genotypes possess Type 2 stability if their responses over environments are parallel to the mean of all genotypes in the trial. Statistics showing Type 2 stability are based on the contribution of each genotype to the GxE variance and linear regression where the slope is equal to 1. A limitation of Type 2 stability is that it is relative to other entries in the trial.

Type 3 stability is defined as small deviations from the regression line modeling genotypic response across environments. An assumption of Type 3 stability is that strict linear response across environments is desirable. As discussed below, the desirability of this type of response is questionable. Lin et al. (1986) strongly criticized Type 3 stability as a descriptive rather than as a predictive model. In particular, the independent variable (that is, environment) usually cannot be measured before the experiment. Type 3 stability would be acceptable in growth chamber experiments where actual environmental factors are known before the experiment.

GxE can be partitioned into variation associated with site (fixed variation) and variation associated with yearly fluctuations within a location (random variation). The within-location interaction is Type 4 stability. Genotypes with less variation at a location over years are judged to be more stable (Lin and Binns, 1988a). Type 4 stability is actually a subset of Type 1 stability that examines the interaction from random events, primarily weather and disease epiphytotics where data from both years and locations are available. Type 1 stability combines both random and fixed variation in its estimate of stability. Type 4 stability has desirable characteristics in that it represents primarily the random portion of the environment and therefore may be applicable to trials over wide regions. It is also independent of the entries in the trial. Lin and Binns (1988a) presented limited evidence that Type 3 stability was probably not an inherited parameter. In contrast, Type 4 stability may be a genetic parameter and therefore may be amenable to selection. Growers may prefer Type 4, or within location, stability (Barah et al., 1981).

Fixed variation that contributes to the environmental variation is the complement to Type 4 stability. Lin and Binns (1988a) did not classify stability across fixed environmental variation. For this chapter, however, the GxE response across fixed variation is termed Type 5 stability. Lin and Binns (1988a) described *site* as the source of fixed environmental variation. Specific cropping systems can be also considered as fixed environmental effects. Type 5 stability may have important implications for crop improvement in sustainable agriculture. It is virtually impossible to develop a specific cultivar for each crop rotation or intercrop combination that may occur in ecoagriculture management systems. Therefore, the development of genotypes that are reliably productive across a range of management schemes is important if true cropping diversity tailored to local environments is ever to be achieved.

Interpretation of GxE through Biometric Models

There are two basic approaches to the analysis of GxE. One can be termed *avoidance* where GxE is removed through partitioning the data into homogeneous groups, or by transforming the data to remove interaction (Freeman, 1985). Where the interaction is nonadditive (that is, responses of genotypes differ in magnitude but not in rank over environments), transformation should be effective. Although these processes may be statistically correct, they complicate biological interpretation. Partitioning of the data may be effective in removing both nonadditive and crossover effects. Crossover interaction is defined as genotypic response differing in rank over environments (Baker, 1988). Cluster analysis is a common method of partitioning test sites to limit crossover rankings (such as, Abou-El-Fittouh et al., 1969).

A second approach is to analyze and understand GxE. Plant breeders have traditionally worked with large numbers of unknown genotypes in ill-defined environments resulting in little understanding of GxE. This empirical approach has led to the release of superior cultivars without much understanding of the reasons for their superiority (Eisemann et al., 1990). Commonly used parametric statistics, nonparametric ranking methods, and regression techniques contribute to a lack of understanding of GxE because they attempt to distill multidimensional phenomena into a single dimension statistic. A systematic analytical approach should elucidate a biological basis for GxE. Such an approach, based on multivariate techniques, may be particularly important for sustainable agriculture because of the complexity of interactions that are possible in such a system. For example, breeding for a multiple cropping system requires the consideration of interspecies interactions that are not present in most monoculture situations. Perhaps the ideal analytical approach would be to possess an understanding at the level of gene by environmental variable interaction. This ideal may eventually be realized by coupling statistical methodology with saturated genome maps (Barry and Geng, 1990). Using this methodology, quantitative trait loci could be identified that are important for determining phenotype in one environment but not another. However, for most breeding programs the point of diminishing returns is quickly reached where the information gained may not be worth the extra effort exerted.

CLASSIFICATION OF G×E MODELS

In subsequent sections, various statistical techniques applicable to GxE analysis will be contrasted and compared. Fig. 10.1 qualitatively depicts the relationship among various GxE models with regard to complexity and information content. Complexity, shown on the ordinate, is a measure of the knowledge required and the physical difficulty in implementing a particular algorithm. The amount of information obtained from a particular analysis is shown on the abscissa. Examples of an algorithm that requires only basic statistical knowledge, and can be done by hand or on a calculator are non-parametric rank sum models. These models impart little information beyond relative stability of genotypes. On the other hand, pattern analysis algo-

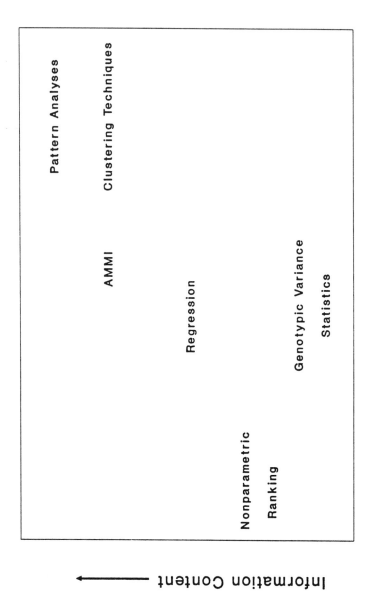

Fig. 10.1. Complexity and information content of some models used in the analysis of GxE interaction.

rithms require specialized statistical knowledge and have only become feasible with the availability of the microcomputer. These models can be considerably more informative than other models because the data from all variables (genotype, environment, and attribute) are simultaneously considered. Most GxE models lie between these two extremes. Some models, such as AMMI (Additive Main effects Multiplicative Interactions), could be termed "efficient" in that they extract a large amount of information with moderately simple algorithms.

Univariate Analysis of GxE Interactions

OVERVIEW

Stability models based on univariate statistics were the first methods developed to measure crop performance across environments (Yates and Cochran, 1938). These methods are based on Gaussian distribution theory and differ from the statistical assumptions that apply to multivariate or nonparametric measures of crop stability. Univariate methods were initially developed for intensive cropping systems; more recent applications to sustainable cropping systems have expanded the use of stability analyses as tools to understand complicated biological systems such as polycultures. Because of the simplicity and flexibility of these methods, they can augment analyses commonly employed to evaluate crop stability, select among superior genotypes, or choose among diverse cropping systems.

Univariate stability measures—more than a dozen—are divided about equally into two broad categories that use either variance or regression methods (Lin et al., 1986). All methods essentially attempt to determine a genotypic response across a random set of environments and anticipate future behavior. A stability response is more accurately measured as the phenotypic change across a specified range of a stress if that stress defines the responses of genotypes (see Drought Susceptibility Index, below, Blum et al., 1989). Multiple undefined stresses, however, often act singularly and in combination to define the plant phenotype in a unique interaction with the genotype. A cultivar must then be tested across a range of environments to define stability.

VARIANCE METHODS

Perhaps the most straightforward analysis of genotypic stability is based on the deviation of the individual phenotype from the average genotype and is computed with variances and the coefficients of variation (CV) (Francis and Kannenberg, 1978). Genotypes with small variances and CVs across environments are biologically stable (Type 1 or Type 4). Stability measures derived from these statistics are easily computed and interpreted within the context of common statistical methodologies. Joint selection for a reduced phenotypic variance and improved genotypic mean is a risk-aversion or safety-first approach that may have greatest use in highly unstable cropping environments. Within this context, selection against variance assumes that variation associated with the environment (σ^2_e) and GxE (σ^2_{gxe}) is reduced rather than genetic variance (σ^2_g). Breeding methodologies for crop stability have typically not addressed the loss of genetic variance relative to environmentally derived variances.

A second group of univariate stability equations is based on GxE interaction variances. These include two Plaisted models (Plaisted and Peterson, 1959; Plaisted, 1960), ecovalence (W^2_i) (Wricke, 1962), and stability variance (σ^2_i) (Shukla, 1972). With these measures, genotypic stability is defined by a response across environments that parallels the mean response of all genotypes under consideration (Type 2 stability) (Lin et al., 1986). Stability is also a function of the diversity and number of both entries and sites in the trial. This, however, is a restriction that applies throughout GxE analysis. A distinct advantage of agronomic stability over biological stability is that genotypes that respond to improved environment or cropping systems are not penalized by the model.

Among this group of stability parameters there are features that differentiate each, but all are related mathematically (Lin et al., 1986). The literature supports a preference for the use of ecovalence and stability variance (Becker and Leon, 1988; Gravois et al., 1990). The stability variance provides an unbiased estimate of GxE variance for each genotype that can be tested for significance and adjusted for covariates. Becker and Leon (1988) described the linear relationship between ecovalence and stability variance and noted that genotypic rank across both measures should remain constant.

The Plaisted methods are based on comparisons of pair-wise GxE variances computed for all combinations (Plaisted and Peterson, 1959) and GxE variance components, where each genotype is jackknifed (Sokal and Rolf, 1981) from the data and subsets of i-1 variances computed (Plaisted, 1960). Although these methods are computationally cumbersome, they are useful to breeders attempting to make a final choice among a few elite breeding lines.

REGRESSION STABILITY MODELS

Regression techniques used to develop stability parameters are based on linear slope and deviation from that slope. Stringfield and Salter (1934) and Yates and Cochran (1938) first applied the joint regression method to stability analysis. Finlay and Wilkinson (1963) popularized joint regression in a study of barley genotypes in Australia. The stability of a genotype was determined by the regression of genotypic means at each site against the mean of all genotypes at each site (environmental index). Regression coefficients of $b=1.0$ indicated a genotypic response parallel to the environmental index. Finlay and Wilkinson (1963) suggested that slopes $b<1.0$ indicated better adaptation to poor environments, while genotypes with $b>1.0$ are best used in superior environments. Crossa (1988) examined the stability of maize genotypes across international test sites and classified the genotypes as responsive ($b>1$) and appropriate for high yield sites or stable ($b<1$) and suitable for low yield sites.

A genotype with an ideal stability response across a wide range of environments would have a stability of $b=0$ in poor environments but respond to favorable environments ($b>1$). Fernandez et al. (1989) proposed segmented linear regression to measure this type of stability and were able to identify a group of mungbean genotypes that fit this definition. One genotype in particular had been identified as high yielding in most environments but was designated as unstable by conventional regression analysis. It is obvious that both Type 2 and Type 3 stability parameters failed to identify this high-yielding, highly stable group because the stability response over environments was nonlinear.

Eberhart and Russell (1966) used the Finlay-Wilkinson joint regression model and attached importance to the R^2 value of each genotype's regression. They assumed a negative relationship between yield and stability in the maize hybrids studied. Their compromise was to select for a genotype with an average regression and a high R^2 value (Type 3 stability). This selection criterion identifies genotypes that perform similarly to the average of the other genotypes. This may be a poor choice of methodology for sustainable agriculture breeding programs. Few genotypes used for composition of the environmental means may have environmental responses in a sustainable system that are worth mimicking.

Perkins and Jinks (1968) varied the Finlay-Wilkinson joint regression model by regressing each genotype's GxE component onto the environmental index. This is analogous to regressing the unsquared ecovalence of a genotype within each environment (Wricke, 1962) onto the index (mean yield) for that environment.

Regression methods have been criticized for their violation of the basic assumption of independence among the x and y variables (Freeman and Perkins, 1971). This problem is minimized as the number of entries and locations become large (Becker and Leon, 1988) or is corrected by the exclusion of the regressed genotype from the environmental index (Freeman and Perkins, 1971). Freeman and Perkins also suggested that a set of test genotypes be used to compute environmental indexes. Considerable thought should also be given to the genotypes that compose the test set and how they relate to the test environments (Brown et al., 1983). Perhaps the most severe criticism is that unless one environmental factor dominates at all sites, the environmental index may contain spurious information. For example, two sites may have identical mean yields but differ ecologically, requiring different sets of genes for adaptation. To attempt to restrict sites to only those that have linearly related environmental indexes may conflict with the statistical requirement of large numbers of test environments and genotypes to provide x and y independence. Other criticisms of regression methods include the positive covariances between mean yield and slope, and the heteroscedasticity (or heterogeneity) of error variances among sites (Lin et al., 1986; Becker and Leon, 1988; Gray, 1982).

Mean trial yield is generally used as the environmental index against which a line's site yield is compared. Mean trial yield is used because it summarizes the multitude of environmental variables that affect plant growth and development. It may not always be the most appropriate index. Rao and Willey (1979) used the total productivity of intercrop systems involving different ratios of pigeonpea and sorghum. Hildebrand (1984) proposed using mean yields of growers as an index to which cultivar performance may be compared. Grower averages are possibly the best estimate of the expected on-farm performance. Experimental trials often measure a potential performance rather than the realized performance of growers. Basing the environmental index on grower averages rather than trial averages may limit bias in stability measures from extension and on-farm yield trials.

Measures of environmental index may include planting density or other continuous agronomic variables. Such a management measure would be used to estimate stability to fixed cropping variables. In reality, it is beyond the budgets of most programs to conduct extensive management trials. As a result, a breeding line may be grown under two conditions representing the extreme conditions of a single important management variable. The phenotypic change between the two extremes is used as a stress sensitivity index. The literature for using this fixed model is probably best defined for drought tolerance. Fischer and Maurer (1978) developed a drought susceptibility index (DSI) to measure relative stability across moisture regimes.

$$\text{DSI} = (1 - Y_{id}/Y_{iw})/(1 - Y_{.d}/Y_{.w})$$

The DSI is a ratio of the yield of genotype i in drought conditions (Y_{id}) to the yield of the same genotype in well watered or unstressed conditions (Y_{iw}). This ratio is standardized by the mean yield of all genotypes in drought $(Y_{.d})$ and well-watered conditions $(Y_{.w})$, where the dot notation signifies summation across the subscripted variable. The DSI has been found to be highly correlated to Type 1 stability as measured by Finlay-Wilkinson joint regression if moisture limitation is the primary factor determining environmental yield variation (Sinha et al., 1986). The DSI has found broad application in a number of crops for identifying genotypes with stability in moisture-limited environments (Chaudhuri et al., 1986; Hatfield et al., 1987; Ehdaie et al., 1988; Blum et al., 1989; Stark et al., 1991). Although developed for measuring drought tolerance, the index would also be useful for measuring tolerance to other biotic and abiotic stresses. The primary advantage of using a susceptibility index over a stability index is cost. A stability index requires sampling a large number of environments to assess interactions with random variation. A susceptibility index requires only sufficient testing to assess the differences among genotypes in response to a fixed effect. The latter is a much simpler proposition than the former. The drawback is that environmental factors other than the specifically measured stress may be important for determining GxE and may not necessarily be addressed by a susceptibility index.

Multivariate Analysis of GxE Interactions

RATIONALE FOR A MULTIVARIATE APPROACH

Yield trials would be of no predictive use if each time a field was planted a completely new order of relative yields of cultivars was observed. In fact, past performance of genotypes in testing programs is predictive of future performance in target environments. There are patterns of GxE variation. The Finlay-Wilkinson joint-regression model is sufficient to explain GxE if there is a single pattern of interaction. In the worst case, there would be as many patterns of variation as there are combinations of genotypes and environments being evaluated. Reality typically falls somewhere in between these two extremes. Multivariate analysis is an attempt to reduce the patterns of variation into a limited set of patterns that can be used by a breeder to understand the observed phenotypic variation.

Multivariate genotypic analysis is an extension of univariate analysis and typically requires data with a multivariate Gaussian distribution (rank correlations are the exception). Multivariate analyses are complex and typically require a programmable off-the-shelf statistical package or a specialty program written for a specific analysis.

CORRELATION ANALYSIS

The oldest multivariate analysis in crop improvement is correlation analysis. Huehn (1990) felt that one of the most important functions of GxE analyses was identification of locations or types of environments that produce correlated cultivar performance. Typically within a breeding program there resides a body of common wisdom as to which locations or cropping methods produce similar genotype rankings. Analysis of the correlation of breeding line phenotypic means or rankings between environments can be used to verify this type of observational data. Breeding crops for low-input environments presents new breeding problems because data previously accumulated for genotypes in high-input conditions may not be predictive of performance in low-input cropping patterns (Atlin and Frey, 1990). Correlation analysis of genotypes in high-input cropping systems with low-input cropping systems or various polyculture trials would suggest the types of testing and selection systems needed for selecting stable low-input cultivars.

CLUSTER ANALYSIS

Cluster analysis is an extension of environmental correlations that uses the similarity of relative phenotypic performance as a measure of environmental similarity. Cluster analyses such as Unweighted Pair Group Method–Average (UPGMA) sequentially joins into larger groups environments with the most similar patterns of genotype performance. It is common for cluster analysis programs to be based on an $n \times n$ correlation table (SAS, 1985) that was in turn formed from a data set of g columns of phenotypic values for each breeding line and n rows of environments where those breeding lines were tested. For a large number of the environments, cluster analysis can render a correlation table into an outline of environments with uniform rankings. Abou-El-Fittouh et al. (1969) and Fox and Rosielle (1982a) provide examples of cluster analysis in GxE interpretation. Canonical or principle component analysis (PCA) can also be applied to an $n \times n$ correlation table as methods of identifying environments that rank genotypes similarly (Byth et al., 1976). Other studies that have used multivariate analysis to group environments of uniform information content include Gusmao et al. (1989) and Peterson and Pfeiffer (1989). The endpoint of these studies is to either pool information across locations for better predictability of genotypic performance or to reduce the amount of testing required to assess a breeding line's GxE.

Cluster analysis can also be used for identification of genotypes with similar patterns of environmental interactions. Fox and Rosielle (1982b) suggested using reference cultivars to which the GxE patterns of new breeding lines are compared. Fox et al. (1990) found, using cluster analysis of international triticale nurseries, that groups of triticale cultivars responded to environments in predictable patterns based on genome composition (complete vs. substituted triticales). There is an emphasis in sustainable agriculture research on using grower-based information (Geng et al., 1990). A cultivar with broad long-term acceptance among growers may well have interaction patterns acceptable to growers. Cluster analysis and pattern analysis are possible methods of identifying new breeding lines with stability patterns similar to cultivars well accepted by growers.

BEYOND CLUSTER ANALYSIS

More complex extensions of multivariate analysis include AMMI (Gauch, 1990; Gauch and Zobel, 1989; Zobel, 1990) and pattern analysis (Byth et al., 1976; Crossa, 1988). The AMMI model is a recent hybrid of the GxE variance methods combining ANOVA (analysis of variance) to analyze main effects (genotypes and environments) and PCA (principal components analysis) methodology to analyze the GxE or multiplicative effects. The univariate approach is appropriate to partition and differentiate among genotypes (main effects), but the standard ANOVA is unable to analyze GxE interaction variance and can only partition it from the main effects of genotype and environmental means. By applying PCA to the GxE variance breeders can more accurately distinguish between random error and unique patterns associated with GxE interactions. A biplot with the first PCA scores plotted against their respective genotypic and environmental means is used to display GxE effects. AMMI analysis yields information about Type 1 stability. A genotype whose PCA score is close to zero shows little interaction with environments. Conversely, adaptation to specific environments is indicated when genotypes and environments possess the same sign. Cartesian proximity of a genotype to an environment indicates the degree of adaptation to that environment. A stability statistic could be devised for

AMMI, where the absolute value for scores of a genotype could be summed over each PCA axis included in the model. Small values would be associated with greater Type 1 stability.

GxE responses can be readily related to an environmental trend or physiological response to the environment using AMMI. Gauch and Zobel (1988) used this technique to distinguish among maturity groups of soybean grown in New York environments, while Wallace and Massaya (1988) used AMMI to gain an understanding of the effects of temperature and photoperiod on the flowering time of the common bean grown in Central America. Environmental and geographic factors may also be associated with AMMI-partitioned GxE variables. As an example, Shafii et al. (1992) used AMMI to analyze GxE for regional rapeseed trials. For both yield and oil content, a gradient was observed on the first PCA axis where northern sites were positively associated with scores of one sign while southern sites were associated with scores of the opposite sign. On the second PCA axis, no clear relationship was observed for yield but an east-west gradient was observed for oil content. The underlying environmental variables with the greatest contribution for each axis were not identified, however.

Basford et al. (1990) proposed a higher dimensional extension to GxE analysis that uses a three-way, three-mode analysis. The multivariate analyses discussed above can analyze the interaction of genotypes with environments, cluster genotypes on the basis of similarity of traits (yield, height, maturity, pedigree, biochemical marker similarity) or group genotypes by similarity of responses to environments. Using three-way, three-mode analysis, three classes of variables (genotypes, environments, plant characters) can be simultaneously considered. This allows the identification of correlations in phenotypic stability between traits or between genotypes for an array of traits such as simultaneously considering stability for both yield and protein. Although Basford et al. (1990) considered only three suites of variables, there is no restriction on the number of variables that can be considered once univariate analysis is abandoned. Theoretically, variables such as weather, fertility, and cropping history could also be included in higher dimensional models.

Nonparametric Measures of Stability

Certain assumptions are made about the evaluated variable within parametric GxE models and include assumptions of uniform and Gaussian distribution of observations in all environments and genotypes. These assumptions may be approached for continuous data such as yield through either luck or transformation. It is rare that traits that are evaluated as a percentage or by visual scoring can be forced to conform to the assumptions underlying univariate stability models. Examples of data having these distribution problems include lodging scores, disease and insect damage scores, seedling emergence or survival, weed competition scores, and visual or mechanical quality scores.

Ranking genotypes within environments for relative performance is a common method of rating genotypes that is indifferent to scale and tolerant of outlying observations. Rank scales are also independent of environment. The stability of ranking, therefore, has properties of distribution and robustness that are desirable for some data. The tradeoff in using a simple measure with fewer assumptions is that some of the resolution available in a parametric test is lost with a rank-based model. Ranks are unable to measure the magnitude of differences between observations. Ranks and other nonparametric GxE models, however, may be more intuitive than some of the more elaborate parametric models. For many breeders rankings are common and familiar data summary techniques.

An example of a nonparametric stability statistic was presented by Huehn (1979). The stability measure $S^{(2)}$ is a measure of the variance of a cultivar's rank across a set of environments. The rank variance is estimated by the model:

$$S_i^{(2)} = \Sigma(r_{ik} - r_{i.})^2/(n-1), \text{ where } r_{i.} = \Sigma r_{ik}/n$$

For this model r_{ik} is the rank of cultivar i in environment k, and n represents the number of environments in which the cultivars were evaluated. The rank $r_{i.}$ then is the mean rank for cultivar i. $S^{(2)}$ then measures the deviations in rank for cultivar i standardized by the number of environments in which it is tested. Nassar and Huehn (1987) developed an approximate chi-square distribution for $S^{(2)}$ that allows the testing of genotypes for the assumption of a common stability value.

Unscaled data will present genotypes that are consistently ranked at the top of trial as having stable yield performance. Nassar and Huehn (1987) also propose scaling the phenotypic values before ranking by adjusting for mean cultivar effects. The rank-by-environment interaction would be based on x_{ik}^*, which is the phenotypic value of cultivar i within an environment k corrected by subtracting the genotypic effect of the cultivar: $x_{i.} - x_{..}$ where $x_{..}$ is the overall mean. Additional rank stability models were presented by Nassar and Huehn (1987) and Huehn (1979) with different distributional properties from the rank variance $S^{(2)}$. If different cultivars with better yield potential were included, a genotype previously measured as stable would then be measured as unstable. Becker and Leon (1988) concluded from previously published work (Nassar and Huehn, 1987) that significant differences in rank stability are commonly reduced by scaling yield before ranking. If a risk-aversion type of stability is preferred, a rank based on consistently good performance at the top of the trials should be chosen.

Huehn's $S^{(2)}$ represents a precise and powerful measurement of cultivar stability. Nonparametric data analysis has the potential to reduce complex data into intuitive measures of stability. For example, Fox et al. (1990), using international triticale yield trials, proposed stratified rankings to define cultivar stability. Cultivars that consistently yield within the top third of trials would be desirable for stability and yield, those consistently within the lower third would be stable but undesirable, and cultivars with yield rankings throughout the distributions would be unstable. The stratified ranking lacks a statistical test of significance but provides a simple summary for identifying cultivars with desirable GxE responses.

A typical problem in cultivar release is the relative reliability of a new cultivar relative to an existing cultivar used by growers. One measure of reliability is the pair-wise rank superiority proposed by Souza and Sunderman (1992), the frequency p of trials in which a new genotype had a superior rank to the reference cultivar. As an example, winter survival data is generally difficult to analyze because of the variability of stress and plant stand. This is a particularly difficult problem for the winter-wheat growing area of the U.S. intermountain region where snowmolds can decimate stands of wheats. Using regular analysis of variance, it was impossible to determine if

'Survivor' (a cultivar with improved snowmold tolerance) had significantly more reliable stands than 'Blizzard' (a cultivar with moderate snowmold tolerance) after eight years of trials (Souza et al., 1989). 'Survivor' consistently—across trials, years, and sites—had better spring stands than 'Blizzard' or 'Manning', another widely accepted winter wheat in the intermountain growing region. 'Survivor' had a superior ranking to 'Blizzard' in 80% of the 22 trials in which the two cultivars were grown together and was superior to 'Manning' in 79% of 26 trials (Souza and Sunderman, 1992). The test of significance for the pair-wise rank superiority uses binomial probability; for the comparison of 'Survivor' and 'Blizzard' the standard error of $p=0.80$ is:

$$(p(1-p)/n)^{0.5} \text{ or } ((0.8 \times 0.2)/22)^{0.5}=0.085$$

If 'Survivor' and 'Blizzard' were equal in potential for establishing a spring stand, in half of the trials 'Survivor' would have a superior rank and in the other half 'Blizzard' would have a superior rank ($p=0.5$). Using a 21-degree of freedom t-value, it is possible to confirm that a p of 0.80 is significantly greater than the null hypothesis at the 99% confidence level. For extension work and preliminary data analysis, rank superiority measures coupled with genotypic means provide simple and accurate representations of GxE interaction and relative desirability of one cultivar over another.

Eskridge (1990) proposed application of safety-first statistics to the problem of measuring the stability of a phenotype across environments, such that the cultivar with the lowest frequency of failure has the most desirable GxE. In a sustainable cropping system, one of the goals of a crop improvement program is to develop cultivars that produce reliably under a broad range of conditions. Disastrous failure as a result of the performance of a cultivar should be minimized. One of the prime problems with a safety-first approach is defining failure. A GxE model based on frequency of success and failure, however, lends itself to analysis using a binomial distribution. Returning to the winter-wheat survival example above, failure may be defined as spring stands that are so poor that winter wheat must be reseeded

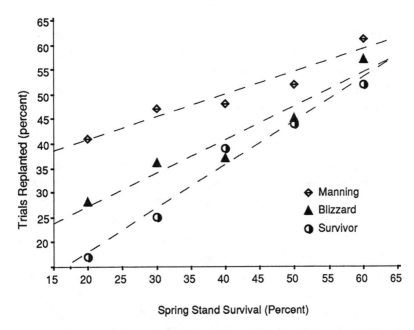

Fig. 10.2. Comparison of three hard red winter wheat cultivars for the frequency of trials with spring stands less than the grower-specified percentage.

with spring wheat. If the decision to reseed is made when spring stands are below 20%, then based on spring stand data in southeastern Idaho from 1983 to 1990, fields of 'Survivor' would be reseeded in 16% of the possible environments and 'Blizzard' in 28%. If, however, the decision to reseed is made when spring stands are below 50% survival the frequency of failure is almost identical for 'Survivor' and 'Blizzard' (Fig. 10.2). A data summary such as Fig. 10.2 is useful for presenting genotype-by-failure threshold interaction to the grower. This allows growers to set their own threshold for failure and make the appropriate varietal selection for their farming practices.

Stability Models for Intercropping

Evaluation of the stability of Land Equivalent Ratios (LERs) is one application of stability models to intercropping systems. LER is a linear combination of "relative yields" summed over the number of species intercropped together (Willey and Osiru, 1972). Relative yield of any one crop in a two-species model is defined by a ratio of the yield of Species A in the presence of Species B divided by the yield of Species A in monoculture. For a two-crop model the LER is defined as:

$$\text{LER} = [A_{i(j)}/A_m] + [B_{(i)j}/B_m]$$

where A_m and B_m are the monoculture yields of Species A and Species B, respectively; $A_{i(j)}$ is the yield of the ith genotype of Species A intercropped with the jth genotype of Species B; and $B_{(i)j}$ is the yield of the jth genotype of Species B intercropped with the ith genotype of Species A.

If the intercropping system includes more than two species, the LER is a linear combination of all relative yield ratios, with components equal to the number of species intercropped. The numerator of each relative yield component would reflect the increased number of species, so that $A_{i(jkl...z)}$ is the yield of the ith genotype of Species A in the presence of the jth, kth, lth . . . zth genotypes of Species B, C, D, and so on, respectively.

LER is most easily analyzed with univariate statistics. One difficulty associated with the relative yield ratio is autocorrelation between the numerator and denominator; however, modifications resolving this problem are presented by Mead and Riley (1981), Mead and Willey (1980), and Federer (1993 and this volume, Chapter 11). They propose that the monocrop yields used in the denominator be based on experiment grand means (that is, A . . .), regional yields external to the experiment (that is, A_e), or optimal yields (A_o). Regardless of the equation used, LER > 1.0 indicates a land-use efficiency of intercropping over monocropping. LER values of 1.0 indicate land-use equivalency between intercropping and monocropping systems.

Wricke's (1962) ecovalence model is defined for monocrops as:

$$W^{2i} = (X_{ik} - \bar{X}_{i.} - \bar{X}_{..})^2$$

where:

X_{ik} is the yield of the ith genotype in the kth environment;

$\bar{X}_{i.}$ and $\bar{X}_{.k}$ are the marginal means of the ith genotype across all environments and the kth environment across all genotypes in the experiment, respectively; and $\bar{X}_{..}$ is the grand mean.

Ecovalence values range from zero upward, with small values indicating high ecovalence (Weber and Wricke, 1990). This model could be adapted to Mead and Willey's (1980) two-species LER equation with monocrop grand means used for the denominator, such that:

$$L^*_{ijk} = (A_{i(j)k}/A_{...}) - (B_{(i)jk}/B_{...})$$

where: L^*_{ijk} is a modification of the LER equation shown above, with environments (k) included. The ecovalence of L^*_{ijk} is then defined as:

$$W^2_{(ij)} = (L^*_{(ij)k} - L^*_{(ij).} - L^*_{(..)k} - L^*_{(..).})^2$$

where $L^*_{(ij)k}$ replaces X_{ik} in Wricke's (1962) ecovalence equation. Marginal means and grand means are similarly replaced and stability $(W^2_{(ij)})$ is defined for a unique combination of the *ith* genotype of Species A with *jth* genotype in Species B across k environments.

This model defines the stability of a unique i and j combination of genotypes, but certain genotypes may confer greater stability across a wider array of companion species. This situation is analogous to general and specific mixing ability in intercropping systems (Federer, this volume, Chapter 11, and 1993).

In most definitions, the k environments are separated by temporal and/or spatial parameters. If the definition of environment is extended to include both biological and physical parameters, then the genotype of one species helps define the environment of the companion crop. Thus, the *ith* genotype of Species A is grown in an environment defined by the *kth* location and *jth* genotype of Species B. Wricke's (1962) stability model may be applied to the actual yields of genotype i of Species A across environments to estimate general stability such that:

$$W^2_a = \sum_{j=1}^{q} \sum_{k=1}^{p} (X_{i(jk)} - \bar{X}_{.(jk)} - \bar{X}_{.(..)})^2,$$

where W^2_a is the general stability of the *ith* genotype of Species A and all other components are as previously defined. Similarly, a general stability parameter could be estimated for the genotypes in Species B, denoted as W^2_b and summed over i and k. This model can be changed to accommodate different analytical needs, but as the number of species intercropped becomes large so does the analysis. Biplots of $W^2_{(ij)}$ against L^*, and W^2_a against the actual yields of the *ith* genotype of Species A would be useful methods of displaying this data.

Applications of Stability Tests

Researchers have different objectives when modeling and analyzing GxE. A breeder may wish to know in preliminary phases of a breeding program what the effects are of GxE on heritabilities and genetic gain. Advanced lines and newly released cultivars are tested in regional nurseries to characterize cultivar performance and predict suitable growing regions. Finally, processes contributing to yield and adaptation may be clarified through detailed analysis of GxE. It is possible to compare the various models and arrive at some conclusions regarding which may be most appropriate for these situations. An additional breeding interest is environment classification. The objective of environment classification may be to identify those environments allowing maximum expression of a particular trait or those that are most representative of the target growing region. Environment classification is discussed above within the context of multivariate models.

WITHIN PROGRAM APPLICATION

A breeder evaluates many lines during early phases of genotype testing. These lines are usually tested over a limited number of locations and years while selecting traits of high heritability. Locations are usually situated near the breeding nursery, leading to correlations among breeding line rankings across locations. Although individual lines may be tested for two or three years, lines are added to or discarded from the nursery annually. Data resulting from these tests may be balanced within years but not necessarily across years. Breeders require quick and simple tests that may be applied to unbalanced data to eliminate the least stable and lowest yielding lines. Ranking methods satisfy this requirement. Linear regression may also be applied if the assumption of linearity among environmental indexes can be met. This assumption is more likely to be met at this stage of testing, where trials are confined to a localized region. Regression provides more detailed information than ranking because specific response to environments is also measured. Lin and Binns (1988b) have also proposed a method in which balanced data is not required. A "superiority measure," P, is defined as the average deviation (across locations) of a cultivar from its maximum performance. This method combines both mean and stability into a single statistic.

ANALYSIS OF ADAPTATION FOR GENOTYPES NEAR RELEASE

Regional testing of advanced materials requires different considerations for the analysis of GxE. A method is needed that evaluates cultivar response over a wide array of environments. A cultivar should be evaluated for adaptation to specific locations and buffering from random environmental variation. Often the coordinator of regional trials receives summarized data from locations and must work with means rather than replicated data. This restriction lowers the power of variance stability statistics. Again, no one technique stands out as the best model for regional trials. It is difficult to meet the restrictions required for regression, particularly that of linearly related environments. AMMI is probably one of the most informative models and can accommodate limited numbers of missing values (Gauch and Zobel, 1990). AMMI is also appropriate when within-location replication data is unavailable; however, model validation tests determining the appropriate number of PC axes are not possible. Other multivariate techniques are applicable and may be as informative as AMMI but require balanced data sets. At this point in cultivar testing intercropping research trials measuring LER stability are appropriate. Estimation of Type 5 stability for advanced lines is also necessary to predict the reliability of on-farm production.

BASIC STUDIES OF YIELD AND ADAPTATION

Analytic approaches to the interpretation of GxE are hampered by a lack of data concerning environmental variables and yield component data and a lack of methods for directly relating that data to the GxE model. The bottleneck is in collecting and relating relevant data, not in choice of model. For example, AMMI models extract the "pattern" from "noise" and display it along one or more PC axes. It is not always intuitively obvious how

215

the pattern on each axis relates to the real world. Pattern on an axis may be made up of more than one environmental variable, or may be some combination of pattern and noise (see above, multivariate discussion and Shafii et al., 1992). Other multivariate models, particularly three-way, three-mode pattern analysis, exploit all data available for analysis since all traits may be considered simultaneously (Basford et al., 1990). Consideration of multiple aspects of the plant phenotype across environments using three-way, three-mode pattern analysis raises the potential for understanding stability relationships among economic traits.

The analytical approach is the current frontier in the study of GxE. Considerable research effort is needed to determine how to handle unbalanced data, relate quantitative environmental variables to GxE models, and determine the relative importance of genetic and environmental components to GxE.

Selecting Stable Cultivars for Sustainable Agriculture

Selecting cultivars with stable phenotypes may be viewed as an exercise in the accumulation of tolerances to stresses. For example, Qualset and Vogt (1980) concluded that by accumulating resistances to a range of diseases and insects through successive backcrossing programs, stability was improved in a wheat cultivar. Avoidance of losses from disease unarguably will improve the yield and stability of a breeding line. In most of the literature, however, selection for stability implies selection for tolerance to nonspecific biotic or quantitative abiotic stresses.

Stability of a cultivar derives from its ability to buffer itself from stress. Allard and Bradshaw (1964) partitioned cultivar stability into: (1) individual buffering capacity and (2) populational buffering capacity. Sustainable agriculture research must be viewed within cropping systems, therefore, a third component should be considered: (3) the genotype's contribution to system stability. The third level of buffering is largely an issue of genetic diversity and cropping diversity. For example, deployment of numerous cultivars that are individually stable but rely on identical disease resistance genes would ultimately lead to unstable performance of the cropping system. Strategies should be adopted by a breeder to select genotypes with improved buffering capacity at each of the three levels.

SELECTION FOR INDIVIDUAL BUFFERING

The appropriate environment for selection is an important issue within sustainable crop improvement. Should it be a low-input or a high-input environment if the goal is phenotype stability in a sustainable cropping system? Smith et al. (1990) and Atlin and Frey (1990) discuss the relative merits of each, including heritability, gain from selection, and correlated responses. Allen et al. (1978) reviewed testing programs of five different species for sufficiency of the number of replications, locations, and years of testing. One aspect they considered was the selection efficiency in low- and high-yield environments. Their results suggest that the relative merits of selection in low- and high-yield environments may be species dependent, preventing broad conclusions of any kind.

Selecting for adaptation to a broad range of environments may best be accomplished by selecting in multiple environments; however, experimental support for this strategy is somewhat limited. Smith et al. (1990) concluded that the best selection strategy for identifying stable genotypes for a sustainable cropping system was selection under different stresses in multiple environments, whether low or high input. Rosielle and Hamblin (1981) concluded that selection for average performance in both stressed and nonstressed environments was preferable to selection in either environment exclusively. Romagosa and Fox (in press) suggested that the use of multiple selection sites alternating between northern and southern Mexico may have contributed to the widespread adaptation of CIMMYT's (International Maize and Wheat Improvement Center) spring wheat germplasm. Lonquist et al. (1979) proposed convergent-divergent recurrent selection (CDRS), with the goal of broadening the adaptation of the selected population. In CDRS a population would be grown at multiple sites in year one; then selected individuals would be intermated at a single site in year two. The intermated population would then be planted at multiple sites in year three for the initiation of the second cycle of selection. Four cycles of CDRS in maize improved the variance estimate of the population by location interaction stability (Plaisted and Peterson, 1959) by 28% over the original base population and by 77% over the average of two single-cross hybrids used as check cultivars (Lonquist et al., 1979). Definitive conclusions cannot be made about the advantages of multisite selection over single-site selection because this study did not include a control population that was selected at a single site.

One problem with multisite selection was suggested by Jensen (1988). If selection of superior genotypes in one environment advances only genotypes that are inferior in the alternate selection environment, then gain from selection may be severely limited, possibly even negative. One solution to this problem is the use of a range of environments that would produce positively but not perfectly correlated responses to selection. Methodologies for identifying correlated sites (AMMI, rank correlation, clustering) are discussed above. The underlying objective in multisite selection for stable genotypes is to eliminate yield loss by accumulating into a single cultivar genes that will provide protection from the range of stresses to which it will be exposed. By culling inferior genotypes, only those buffered against a broad array of stresses are advanced through multiple cycles of selection.

An alternative to sequential selection is simultaneous selection, such as testing the same line in multiple environments. Brennan and Byth (1979) found that selection for lines with the highest average yield using standardized units was the best method for identifying genotypes with best yield across environments. This type of selection would result in a risk-aversion type of stability. The disadvantage of selection based on multisite averages is that it increases the size of the testing program and forces the breeder to delay selection until sufficient quantities of seed are available for each line.

POPULATION BUFFERING

Identification of the genotype with the best environmental buffering is one component of breeding for stability in a sustainable cropping system. A second is to improve the buffering capacity of the collection of genotypes that comprise a finished cultivar. Theoretically, frequency of alleles conferring susceptibility determines the risk of yield loss from that stress. The presence of heterozygosity and heterogeneity within a cultivar should, therefore, improve the buffering capacity of the population. Methods of improving populational buffering include the use of open pollinated populations rather than hybrids, blends or multilines rather than purelines or single clones, and heterogeneity within cultivars. As an example of this, Eberhart and Russell (1969) concluded that double-cross hybrids had greater Type 3 stability than single-cross hybrids.

Field research attempting to relate genetic diversity within cultivars with phenotypic stability has had mixed results. Marshall and Brown (1973) developed a rigorous methodology for explaining why and when multilines should be more stable than individual components. It is clear from their theoretical approach that multilines or any other blended population are not inevitably more stable than their component lines. Examples where genetic diversity within cultivars was found to be unrelated to stability include Walker and Fehr (1978), Martin and Alexander (1986), and Bowen and Schapaugh (1989). Jensen (1988), summarizing the effects of multiblends versus purelines, concluded that the greatest advantage in multiblends for mean performance and stability of performance was observed in stressed environments. Blends of genotypes and even species are common within subsistence farming systems that require stability (Merrick, 1990). Ayeh (1988) documented that blends of bean landraces were more stable than expected from the stability of the component landraces when grown under rainfed conditions. Frey and Maldonado (1967) studied the response to planting date of 6 purelines and 57 blends derived from the 6 purelines. They found that the yield of the mixtures was more stable across planting dates than that of the purelines. Federer et al. (1982) developed a statistical analysis based on diallel models for general and specific combining ability to identify genotypes that would respond favorably in blends. Effects modeled were termed general and specific mixing ability and were based on a genotype's effect on mean bi-blend (a 50:50 mixture of two genotypes) performance when that genotype is included in the bi-blend (Federer et al., 1982; this volume, Chapter 11). Selection methodology for determining a genotype's contribution to the stability of a blend is poorly developed. The best number of components to include is usually determined to be the optimal balance of yield versus stability of yield. This is because only a very limited number of genotypes is typically available with peak yield potential. The stability of a blend, however, would probably be enhanced by the inclusion of the maximum number of genotypes that would still produce an acceptable mean performance and uniformity (Ayeh, 1988; Leon and Diepenbrock, 1987).

SIMULTANEOUS SELECTION FOR PERFORMANCE AND PERFORMANCE STABILITY

Issues of selection for GxE stability are defined for advanced breeding lines as identification of optimal combinations of performance and stability. Performance (usually yield) and stability, however, may be negatively correlated (Finlay and Wilkinson, 1963; Eberhart and Russell, 1966). This observation led Eberhart and Russell (1966) to hypothesize that the ideal genotype for temperate climate maize would have high yield, one closely predicted by the average performance of all genotypes within each environment, and average stability. Improvements in yield and yield stability may be negatively correlated, but they are not perfectly so. Heinrich et al. (1985) found that grain yield and stability were independent for 56 sorghum hybrids. Yield and yield stability are certainly not mutually exclusive. The goal of selection then should be the identification of unique lines with optimal combinations of performance and stability of performance.

Brennan and Byth (1979) combined mean yield of wheat genotypes with joint regression stability by scaling the absolute value of the stability slope to the same range as the yield values. The difference between the mean yield and scaled stability value formed a selection index where high-yielding stable genotypes had the highest value. Brennan and Byth (1979) in a separate analysis assigned three different weights to the GxE sum of squares for each cultivar to get a range of magnitudes of GxE effects relative to mean yield. Three selection indexes were calculated as the differences between mean genotype yield and each of the three different weighted GxE sums of squares. Selection indexes that included both stability parameters and mean yield in the variable environments of southeastern Queensland were no better at selecting for average yield than simple selection for average yield alone and in several cases were much worse. Selection on the basis of stability alone produced the worst progress for average yield. Average yield may not be the best measure of success for these types of selection indexes. Within a sustainable cropping system it may be useful to have cultivars that guarantee a base level of production in each environment and also have the potential to respond to favorable conditions. Joint selection for yield and stability may be a way of identifying these genotypes.

Kang (1988) proposed rank summation as a method of combining selection for yield and stability. Genotypes would be ranked for yield from highest to lowest and for stability from most stable to least stable. The two ranks would be summed and the genotypes with lowest summed ranks (I^1) would be selected. This method gives yield and stability equal weight, which may not always be the most desirable combination. This method also assumes no correlative relationship between yield and the measure of stability; however, correlation can and does occur (Lin et al., 1986). Kang and Pham (1991) experimented with different relative weights. For example, if the breeder were selecting crops for subsistence farmers with little reserve food or cash, it may be desirable to weight stability more heavily than yield. Pham and Kang (1988) tried weighting yield more heavily than stability by multiplying I^1 by the yield rank ($I^2 = I^1$ x yield rank). They in turn multiplied I^2 by the yield rank again to produce I^3. With each of these indexes the genotype with the lowest rank summation would be selected. In their study of maize hybrids in semitropical environments, Kang and Pham (1991) found that stability rank had little effect on I^3. An ideal range of weights to include the influence of both factors would, therefore, be between I^1 and I^2.

The rank-summation procedure could be simplified by using a ratio of the genotypic mean to ecovalence (X_i / W_i^2) or the stability variance ($X_i / \sigma^2 x_i$). As with rank-summation, genotypes with high means and low GxE variance would rank highest, while high-mean/high-variance and low-mean/low-variance would have an intermediate rank. Genotypes with low-mean/high-variance would rank lowest. Interpretation could be further aided by a biplot of genotypic means (X_i) and GxE variances (W_i^2), which would provide an additional graphic display of these four categories. This sort of approach is similar to the coefficient of variability (CV) but is inverted and uses different variance components (Francis and Kannenberg, 1978). With rank-sum indexes all entries are equally spaced by rank position and interpretations are usually accompanied by discrimination among actual genotype values. Use of genotypic means to GxE variance ratio along with a biplot provides for both numerical and graphic interpretations of yield and stability.

Rank summation and arithmetic ratios are fairly arbitrary weightings. Barah et al. (1981) proposed that the relative merit of yield and stability be weighted by a grower-determined economic utility function. The utility function would reflect the grower's desire for peak yield versus the desire to avert losses. Bands of "iso-utility" based on the grower utility function can be established to measure the tradeoff of yield and stability. This would allow comparisons between cultivars with different yield and yield stability. Barah et al. (1981) found that for sorghum in central India most growers had utility functions that favored the hybrid with the highest average yield in almost every case. This may not necessarily be the result if grower utility was estimated in other environments, particularly ones with more unreliable growing conditions.

Heritability is not included as a consideration in the above trait selections. Yield and stability are weighted as if they were of equal heritability. Measures of GxE response are higher-order statistics and as such phenotypic and genotypic variances are not well defined at present by the literature. In fact, heritability is typically defined for a specific environment. By contrast, a stability measure is an inference about the change in phenotype across environments. A stability parameter may not have a heritability in the traditional definition because of this incongruity. Studies of the variability of stability statistics report their results as repeatability rather than heritability (Eagles and Frey, 1977). Stability measures are not highly reproducible and typically require a large number of site-years to establish an accurate measure of relative stability (Leon and Becker, 1988). Relative ranks of stability may change as well depending on the yield potential of sites where the stability measure is estimated (Crossa, 1988) and on the method used for computing stability (Becker, 1981). Eagles and Frey (1977) found that the repeatability of yield stability measures in oat was poor to nonsignificant for data sets that included 12 environments. Finlay (1971) did find a significant repeatability in joint regression measures of stability. The contrast of Eagles and Frey (1977) and Finlay (1971), testing in entirely different environments, lends support to the dependence of the level of repeatability in statistics being partially dependent on the cohort of test sites used in measuring stability. Fatunla and Frey (1976) found that the repeatability of the joint regression statistic is dependent on genotype as well, with some populations having nonsignificant repeatability and others quite high repeatability.

Equal weighting of a well-defined mean performance and possibly spurious stability measures may result in unneeded discarding of breeding lines with improved performance. The traditional method of simultaneous selection for traits of different heritability is the selection index (Baker, 1986). The selection index has not been widely applied to joint selection for trait mean and stability, possibly because of the complex sampling and distribution properties of most stability measures. Louw (1990) modified the selection index for application to GxE and this remains a promising area for study.

Indirect selection for stability using highly heritable traits that are genetically correlated to stability may be of value. For example, Haugerud and Collinson (1990) cite maturity of maize cultivars as being an important determinant of yield stability for subsistence cropping in western Kenya and Burundi. They found that early maturity was important not only for the maize crop itself but also for stability of subsequent maize or legume crops. In this case, simultaneous selection for yield and early maturity may have the correlated effect of improving yield stability. Other highly heritable traits that may have correlations to yield stability are plant stature and insect and disease resistance. Physiological studies have been successful in identifying a number of moderately heritable traits such as canopy temperature that are positively correlated to drought-related stability (Blum et al., 1989; Stark et al., 1991). Similar studies using susceptibility indexes may be successful in identifying traits that may be correlated to Type 5 stability. The

223

use of higher dimension GxE models such as three-way, three-mode analysis also holds the possibility of identifying correlated traits that would make indirect selection for stability feasible.

Conclusions

Selection for stability in sustainable agriculture systems presents problems similar to stability selection in high-input cropping systems: testing genotypes across a range of target growing conditions, modeling phenotypic responses to environmental changes, and selecting genotypes with an optimum combination of traits and environmental responses.

Issues that separate selection for sustainable agriculture and high-input systems hinge primarily on the environmental index. As discussed in the section on univariate GxE models, the simple use of mean yield as an environmental index has limited use. Identifying the appropriate measure of environment to calibrate genotypic response is a relatively undeveloped area of GxE modeling but critical to its utility (such as Hildebrand, 1984). Development of a representative set of cropping tests to measure Type 5 stability is a challenge to breeders. The genotype selected for high-input cropping systems may be entirely inappropriate under an alternative management system (Atlin and Frey, 1990). In parallel, there is no *a priori* assumption that phenotype stability within one cropping system translates to phenotype stability in a second cropping system. Environmental classification and clustering of cropping systems are appropriate methods for determining the best locations and management practices for stability measurement.

Other major problems facing those wishing to use GxE models are the plethora of GxE models, lack of user-friendly software, lack of clear means for analyzing unbalanced data, and a lack of ways to relate quantitative environmental data to existing GxE models (Romagosa and Fox, 1992). On the basis of utility, descriptive power, and statistical validity, ecovalence, Stress Susceptibility Index, rank superiority or variance, and AMMI stand out as the most applicable models for crop improvement for sustainable agriculture. Improvements in computer software will undoubtedly solve the

224

problems discouraging widespread implementation of these models by crop improvement programs.

A challenge to breeders and geneticists is the integration of selection for stability into the selection of traits of direct economic value. Kloppenberg (1988, p.162–163) noted that the introduction of improved high-yielding cultivars often leads directly to the displacement of stable local cultivars. The use of genetically uniform elite cultivars often leads to cyclic disease and pest epidemics not prevalent with unimproved cultivars. Kloppenberg's point was to address the genetic vulnerability of modern cropping systems, but the causal force of varietal replacement is important to the discussion of GxE in sustainable agriculture. Growers in present intensive agricultural systems prefer (by choice or economic necessity) cultivars that maximize production in the near term despite the potential for phenotypic and cropping system instability. Growers in present nonintensive agricultural systems need alternatives to low-yielding local cultivars and high-yielding, unstable improved cultivars. Breeders cannot expect acceptance of cultivars based on stability alone. Perhaps the best opportunities for improvement may lie in the development of high-yielding cultivars with buffering derived from genetic heterogeneity. A second opportunity may lie with the development of selection indexes providing appropriate weights to stability and average economic return.

References

Abou-El-Fittouh, H.A., J.O. Rawlings, and P.A. Miller. 1969. Classification of environments to control genotype by environment interactions with an application to cotton. Crop Sci. 9:135–140.

Allard, R.W., and A.D. Bradshaw. 1964. Implications of genotype-environment interactions in applied plant breeding. Crop Sci. 4:503–508.

Allen, F.L., R.E. Comstock, and D.C. Rasmusson. 1978. Optimal environments for yield testing. Crop Sci. 18:747–751.

Atlin, G.N., and K.J. Frey. 1990. Breeding crop varieties for low-input agriculture. Amer. J. Altern. Agric. 4:53–58.

Ayeh, E. 1988. Evidence for yield stability in selected landraces of bean (*Phaseolus vulgaris*). Exp. Agric. 24:367–373.

Baker, H.C. 1981. Correlations among some statistical measures of phenotypic stability. Euphytica 30:835–840.

Baker, R.J. 1986. Selection indices in plant breeding. CRC Press, Boca Raton, FL.

Baker, R.J. 1988. Analysis of genotype-environment interactions in crops. ISI Atlas Sci.- Animal & Plant Sci. 1:1–4.

Barah, B.C., H.P. Binswanger, B.S. Rana, and N.G.P. Rao. 1981. The use of risk aversion in plant breeding: concept and application. Euphytica 30:451–458.

Barry, T.A,. and S. Geng. 1990. A simulation approach to analyze and interpret genetic-environmental interaction. p.94–107. *In* M.S. Kang (ed.) Genotype-by-environment interaction and plant breeding. Louisiana State University Agricultural Center, Baton Rouge, LA.

Basford, K.E., P.M. Kroonenberg, and I.H. DeLacy. 1990. Three-way methods for multiattribute genotype by environment data: An illustrated partial survey. p.234–261 *In* M.S. Kang (ed.) Genotype-by-environmental interaction and plant breeding. Louisiana State University Agricultural Center, Baton Rouge, LA.

Becker, H.C. 1981. Correlations among some statistical measures of phenotypic stability. Euphytica 30:835–840.

Becker, H.C., and J. Leon. 1988. Stability analysis in plant breeding. Plant Breeding 101:1–23.

Blum, A., L. Shipler, G. Golan, and J. Mayer. 1989. Yield stability and canopy temperature of wheat genotypes under drought-stress. Field Crops Res. 22:289–296.

Bowen, C.R., and W.T. Schapaugh, Jr. 1989. Relationships among charcoal rot infection, yield, and stability estimates in soybean blends. Crop Sci. 29:42–46.

Brennan, P.S., and D.E. Byth. 1979. Genotype x environment interactions for wheat yields and selection for widely adapted wheat genotypes. Aust. J. Agric. Res. 30:221–232.

Brown, K.D., M.E. Sorrells, and W.R. Coffman. 1983. Methods for classification and evaluation of testing environments. Crop Sci. 23:889–893.

Byth, D.E., R.L. Eisemann, and I.H. DeLacy. 1976. Two-way pattern analysis of a large data set to evaluate genotypic adaptation. Heredity 37:215–230.

Chaudhuri, U.N., M.L. Deaton, E.T. Kanemasu, G.W. Wall, V. Marcarian, and A.K. Dobrenz. 1986. A procedure to select drought-tolerant sorghum and millet genotypes using canopy temperature and vapor pressure deficit. Agron. J. 78:490–494.

Crossa, J. 1988. A comparison of results obtained with two methods for assessing yield stability. Theor. Appl. Genet. 75:460–467.

Eagles, H.A., and K.J. Frey. 1977. Repeatability of the stability-variance parameter in oats. Crop Sci. 17:253–256.

Eberhart, S.A., and W.A. Russell. 1966. Stability parameters for comparing varieties. Crop Sci. 6:36–40.

Eberhart, S.A., and W.A. Russell. 1969. Yield and stability for a 10-line diallel of single-cross and double-cross maize hybrids. Crop Sci. 9:357–361.

Ehdaie, B., J.G. Waines, and A.E. Hall. 1988. Differential responses of landrace and improved spring wheat genotypes to stress environments. Crop Sci. 28:838–842.

Eisemann, R.L., M. Cooper, and D.R. Woodruff. 1990. Beyond the analytical methodology: Better interpretation and exploitation of genotype-by-environment interaction in breeding. p.108–117 In M.S. Kang (ed.) Genotype-by-environment interaction and plant breeding. Louisiana State University Agricultural Center, Baton Rouge, LA.

Eskridge, K.M. 1990. Selection of stable cultivars using a saftey-first rule. Crop Sci. 30:369–374.

Fatunla, T., and K.J. Frey. 1976. Repeatability of regression stability indexes for grain yield of oats (Avena sativa L.). Euphytica 25:21–28.

Federer, W.T. 1993. Statistical design and analysis for intercropping experiments. Vol. 1. Springer-Verlag, New York.

Federer, W.T., J.C. Connigale, J.N. Rutger, and A. Wijesinha. 1982. Statistical analyses of yield from uniblends and biblends of eight dry bean cultivars. Crop Sci. 22:111–114.

Fernandez, G.C.J., H.K. Chen, and J. Creighton Miller, Jr. 1989. Adaptation and environmental sensitivity of mungbean genotypes evaluated in the international nursery. Euphytica 41:253–261.

Finlay, K.W. 1971. Breeding for yield in barley. Int. Barley Genet. Symp. Proc. 2:338–345.

Finlay, K.W., and G.N. Wilkinson. 1963. The analysis of adaptation in a plant-breeding programme. Aust. J. Agric. Res. 14:742–754.

Fischer, R.A., and R. Maurer. 1978. Drought resistance in spring wheat cultivars. I. Grain yield response. Aust. J. Agr. Res. 29:897–907.

Fox, P.N., and A.A. Rosielle. 1982a. Reducing the influence of environmental main-effects on pattern analysis of plant breeding environments. Euphytica 31:645–656.

Fox, P.N., and A.A. Rosielle. 1982b. Reference sets of genotypes and selection for yield in unpredictable environments. Crop. Sci. 22:1171–1175.

Fox, P.N., B. Skovmand, B.K. Thompson, H.J Braun, and R. Cormier. 1990. Yield and adaptation of hexaploid spring triticale. Euphytica 47:57–64.

Francis, T.R., and L.W. Kannenberg. 1978. Yield stability studies in short season maize. I. A descriptive method for grouping genotypes. Can. J. Plant Sci. 58:1029–1034.

Freeman, G.H. 1985. The analysis and interpretation of interactions. J. Appl. Stat. 12:3–10.

Freeman, G.H., and J.M. Perkins. 1971. Environmental and genotype by environmental components of variability: VIII. Relations between genotype grown in different environments and measures of these environments. Heredity 27:15–23.

Frey, K.J., and V. Maldonado. 1967. Relative productivity of homogeneous and heterogeneous oat cultivars in optimum and sub-optimum environments. Crop Sci. 7:532–535.

Gauch, H.G., Jr. 1990. Using interaction to improve yield estimates. p.141–150. In M.S. Kang (ed.) Genotype-by-environment interaction and plant breeding. Louisiana State University Agricultural Center, Baton Rouge, LA.

Gauch, H.G., Jr., and R.W. Zobel. 1990. Imputing missing yield trial data. Theor. Appl. Genet. 79:753–761.

Gauch, H.G., Jr., and R.W. Zobel. 1989. Accuracy and selection success in yield trial analysis. Theor. Appl. Genet. 76:1–10.

Gauch, H.G., Jr., and R.W. Zobel. 1988. Predictive and postdictive success of statistical analyses of yield trials. Theor. Appl. Genet. 76:1–10.

Geng, S., C.E. Hess, and J. Auburn. 1990. Sustainable agriculture systems: Concepts and definitions. J. Agron. Crop Sci. 165:73–85.

Gravois, K.A., K.A.K. Moldenhauer, and P.C. Rohman. 1990. Genotype-by-environment interaction for rice yield and identification of stable, high-yielding genotypes. p.181–188. *In* M.S. Kang (ed.) Genotype-by-environment interaction and plant breeding. Louisiana State University Agricultural Center, Baton Rouge, LA.

Gray, E. 1982. Genotype x environment interactions and stability analysis for forage yield of orchardgrass clones. Crop Sci. 22:19–23.

Gusmao, L., J.T. Mexia, and M.L. Gomes. 1989. Mapping of equipotential zones for cultivar yield pattern evaluation. Plant Breeding 103:293–298.

Haugerud, A., and M.P. Collinson. 1990. Plants, genes, and people: Improving the relevance of plant breeding in Africa. Exp. Agric. 26:341–362.

Hatfield, J.L., J.E. Quisenberry, and R.E. Dilbeck. 1987. Use of canopy temperatures to identify water conservation in cotton germplasm. Crop Sci. 27:269–273.

Heinrich, G.M., C.A. Francis, J.D. Eastin, and M. Saeed. 1985. Mechanisms of yield stability in sorghum. Crop Sci. Vol. 25. 1109–1113.

Hildebrand, P.E. 1984. Modified stability analysis of farmer managed, on-farm trials. Agron. J. 76:271–274.

Huehn, M. 1979. Beitrage zur Erfassung der phanotypischen Stabilitat. I. Voschlag einiger auf Ranginformationen beruhenden Stabilitasparameter. EDP in Medicine and Biology 10:112–117.

Huehn, M. 1990. Nonparametric estimation and testing of genotype x environment interaction by ranks. p.69–93. *In* M.S. Kang (ed.) Genotype-by-environment interaction and plant breeding. Louisiana State University Agricultural Center, Baton Rouge, LA.

Jensen, N.F. 1988. Plant breeding methodology. Wiley, New York, NY.

Kang, M.S. 1988. A rank-sum method for selecting high-yielding, stable corn genotypes. Cereal Res. Comm. 16:113–115.

Kang, M.S. (ed.). 1990. Genotype-by-environment interaction and plant breeding. Louisiana State University Agricultural Center, Baton Rouge, LA.

Kang, M.S., and D.P. Gorman. 1989. Genotype x environment interaction in maize. Agron. J. 81:662–664.

Kang, M.S., and H.M. Pham. 1991. Simultaneous selection for high yielding and stable crop genotypes. Agron. J. 83:161–165.

Kloppenberg, J.R., Jr. 1988. First the seed: The political economy of plant biotechnology. Cambridge University Press, New York, NY.

Leon, J., and H.C. Becker. 1988. Repeatability of some statistical measures of phenotypic stability: Correlations between single year results and multi years results. Plant Breeding 100:137–142.

Leon, J., and W. Diepenbrock. 1987. Yielding ability of pure stands and equal proportion blends of rapeseed (Brassica napus L.) with double-low quality. J. Agron. Crop Sci. 159:82–89.

Lin, C.S., M.R. Binns, and L.P. Lefkovitch. 1986. Stability analysis: Where do we stand? Crop Sci. 26:894–900.

Lin, C.S., and M.R. Binns. 1988a. A method of analyzing cultivar x location x year experiments: A new stability parameter. Theor. Appl. Genet. 76:425–430.

Lin, C.S., and M.R. Binns. 1988b. A method for assessing regional trial data when the test cultivars are unbalanced with respect to locations. Can. J. Plant Sci. 68:1103–1110.

Lonquist J.H., W.A. Compton, J.L. Geadelmann, F.A. Loeffel, B. Shank, and A.F. Troyer. 1979. Convergent-divergent selection for area improvement in maize. Crop Sci. 19:602–604.

Louw, J.H. 1990. A selection index to cope with genotype-environment interaction with an application to wheat breeding. Plant Breeding 104:346–352.

Marshall, D.R., and A.H.D. Brown. 1973. Stability of performance of mixtures and multilines. Euphytica 22:405–412.

Martin, J.M., and W.L. Alexander. 1986. Intergenotypic competition in biblends of spring wheat. Can. J. Plant Sci. 66:871–876.

Mead, R., and J. Riley. 1981. A review of statistical ideas relevant to intercropping research. J. Royal Statistical Soc., Series A 144:462–509.

Mead, R., and R.W. Willey. 1980. The concept of a "land equivalent ratio" and advantages in yield from intercropping. Exp. Agric. 16:217–228.

Merrick, L.C. 1990. Crop genetic diversity and its conservation in traditional agroecosystems. p.3–11. *In* M.A. Altieri and S.B. Hecht (ed.) Agroecology and small farm development. CRC Press, Boca Raton, FL.

Nassar, R., and M. Huehn. 1987. Studies on estimation of phenotypic stability: Tests of significance for nonparametric measures of phenotypic stability. Biometrics 43:45–53.

Perkins, J.M., and J.L. Jinks. 1968. Environmental and genotype-environmental components of variability. III. Multiple lines and crosses. Heredity 23:339–356.

Peterson, C.J., and W.H. Pfeiffer. 1989. International winter wheat evaluation: Relationships among test sites based on cultivar performance. Crop Sci. 29:276–282.

Pham, H.N., and M.S. Kang. 1988. Interrelationships among and repeatability of several stability statistics estimated from international maize trials. Crop Sci. 28:925–928.

Plaisted, R.L. 1960. A shorter method for evaluating the ability of selections to yield consistently over locations. Am. Potato J. 37:166–172.

Plaisted, R.L., and L.C. Peterson. 1959. A technique for evaluating the ability of selections to yield consistently in different locations or seasons. Amer. Potato J. 36:381–385.

Qualset, C.O., and H.E. Vogt. 1980. Efficient methods of population management and utilization in breeding wheat for Mediterranean-type climates. p.166–188. *In* V.A. Johnson (ed.) Proc. Third Int. Wheat Conf., Madrid, Spain. Nebraska Agric. Expt. Sta., Lincoln, NE.

Rao, M.R., and R.W. Willey. 1979. Stability of performance of a pigeonpea/sorghum intercrop system. p.306–317. *In* International Crops Research Institute for the Semi-Arid Tropics (ICRISAT). Proc. Int. Workshop on Intercropping. Jan. 10–13, 1979. ICRISAT, Patancheru, India.

Rasmussen, D.C., and J.W. Lambert. 1961. Variety x environment interactions in barley variety tests. Crop Sci. 1:261–262.

Rosielle, A.A., and J. Hamblin. 1981. Theoretical aspects of selection for yield in stress and non-stress environments. Crop Sci. 21:943–946.

Romagosa, I., and P.N. Fox. (in press). Genotype x environment interaction and adaptation. *In* I. Romagosa and P.N. Fox (ed.) Plant breeding: Principles and prospects. Chapman and Hall, NY.

SAS Inst. Inc. 1985. SAS user's guide: Statistics. Version 5 ed. SAS Inst., Cary, NC.

Shafii, B., K.A. Mahler, W.J. Price, and D.L. Auld. 1992. Genotype by environment interaction effects on yield and oil content of winter rapeseed. Crop Sci. 32 (in press).

Shukla, C.K. 1972. Some statistical aspects of partitioning genotype-environmental components of variability. Heredity 28:237–245.

Sinha, A., P.K. Aggarwal, G.S. Chaturvedi, A.K. Singh, and K. Kailasnathan. 1986. Performance of wheat and triticale cultivars in a variable soil-water environment. I. Grain yield statistics. Field Crops Res. 13:289–299.

Smith, M.E., W.R. Coffman, and T.C. Barker. 1990. Environmental effects on selection under high and low input conditions. p.261–272. *In* M.S. Kang (ed.) Genotype-by-environment interaction. Louisiana State University Agricultural Center, Baton Rouge, LA.

Sokal, R.R., and F.J. Rolf. 1981. Biometry. 2nd ed. W.H. Freeman, New York, NY.

Souza, E., and D.W. Sunderman. 1992. Pairwise rank superiority of winter wheat genotypes for spring stand. Crop Sci. 32(4):938–942.

Souza, E., D.W. Sunderman, and K. Kephart. 1989. Multi-year regression analysis of snow mold tolerance in winter wheat. p.100. *In* Agronomy abstracts. Am. Soc. Agron., Madison, WI.

Stark, J.C., J.J. Pavek, and I.R. McCann. 1991. Using canopy temperature measurements to evaluate drought tolerance of potato genotypes. J. Amer. Soc. Hort. Sci. 116:412–415.

Stringfield, G.H., and R.M. Salter. 1934. Differential responses of corn varieties to fertility levels and seasons. J. Agric. Res. 49:991–1000.

Unander, D.W., R. Diaz-Donaire, J.S. Beaver, J. Cern-Garcia, and D. Rueda. 1989. Yield stability of dry bean genotypes in Honduras. J. Agric. Univ. Puerto Rico 73:339–347.

Wallace, D.H., and P.N. Massaya. 1988. Using yield trial data to analyze the physiological adaptation of yield accumulation and the genotype by environment interaction effect on yield. Ann. Rept. Bean Impr. Coop. 31:7–24.

Walker, A.K., and W.R. Fehr. 1978. Yield stability of soybean mixtures and multiple pure stands. Crop Sci. 18:719–723.

Weber, W.E., and G. Wricke. 1990. Genotype x environment interaction and its implications in plant breeding. p.1–19. In M.S. Kang (ed.) Genotype-by-environment interaction and plant breeding. Louisiana State University Agricultural Center, Baton Rouge, LA.

Willey, R.W., and D.S.O. Osiru. 1972. Studies on mixtures of maize and beans with particular reference to plant populations. J. Agric. Sci. 79:517–529.

Wricke, G. 1962. Uber eine methods zur erfasung der okologischen streu-breite in feldversuchen. Z. Planzenzuecht. 47:92–96.

Yates, F., and W.G. Cochran. 1938. The analyses of groups of experiments. J. Agric. Sci. 28:556–580.

Zobel, R.W. 1990. A powerful statistical model for understanding geno-type-by-environment interaction. p.126–140. In M.S. Kang (ed.) Genotype-by-environment interaction and plant breeding. Louisiana State University Agricultural Center, Baton Rouge, LA.

11

Statistical Design and Analysis of Intercropping Experiments

Walter T. Federer

The terms *intercropping, multiple cropping,* and *polycropping* are considered synonymous for the purposes of this paper. Statistical design covers the many aspects of planning and designing experiments (Federer, 1984). Three particular aspects of statistical design are addressed in this paper, that is, treatment design, experiment design, and experimental unit (eu) technique, with most of the emphasis being placed upon statistical analyses for the various experiment and treatment designs.

Experiment design, the *arrangement* of treatments (entities of interest) in an experiment, for intercropping experiments does not depend upon whether or not the experiment involves mixtures of crops or lines or only sole or single-crop treatments. The selection of an experiment design depends upon the type and nature of experimental variation in the place where the experiment is to be conducted. Blocking and confounding considerations and *not* types of treatment are the factors determining which experiment design is selected for the experiment.

Treatment design, the *selection* of entities to be included in an ex-

This chapter is Paper No. BU-1115-MA in the Technical Report Series of the Biometrics Unit, Cornell University, Ithaca, NY.

Editor's Note: This chapter is focused on intercropping, but many of the principles apply to stress conditions caused by lower-input cropping approaches.

periment, is crucially connected with the type of statistical analysis appropriate for an experiment. The nature and type of statistical analysis is, or should be, determined by the treatment design. It should not then be a surprise to find that statistical analyses for intercropping experiments are usually vastly different from those used for sole cropping. Also, rather than performing a single statistical analysis for a characteristic as is usually done for a sole cropping experiment, several varied analyses will be necessary to elicit the information from data obtained from an intercropping experiment. Each intercropping experiment usually is approached from several different angles. The following represent some aspects of interest to an experimenter:

- Land utilization.
- Economic or other value.
- Nutrition.
- Multivariate analysis.
- Soil erosion and structure.
- Density and intimacy.
- Biological modeling aspects.
- Sustainability of cultivar yields to meet population needs.

Each of the above multiple scientific and practical aspects of an intercropping experiment, along with the associated statistical procedures, will be briefly discussed in the following sections.

Experiment Design

In planning and designing experiments, the five axioms presented by Federer (1984) should be followed in evaluating the performance of lines from a plant breeding program or entities from other programs. The choice of an experiment design, a plan for the arrangement of treatments in an experiment, is crucial in controlling experimental heterogeneity among the experimental units (eu's), the smallest unit to which *one* treatment is applied. There are many principles for experiment designs (Federer, 1984), but three of the chief ones for an experimenter are randomization, blocking or stratification, and confounding. Randomization assures fair (unbiased) treatment

235

comparisons and estimations of an error mean square; grouping or blocking allows control or elimination of heterogeneity among eu's by grouping like eu's, minimizing variation among eu's within a group, and by maximizing differences among groups (blocks or strata). Partial confounding of treatment comparisons allows the use of smaller blocks that may be necessary to control experimental variation.

Complete block or incomplete block design may be used. From the published literature on intercropping experiments, most are designed as randomized complete block designs and a relatively small number as completely randomized, split plot, or split split plot designs. From a cursory review of the literature, blocking did not control as much of the variation as it should, resulting in fairly high coefficients of variation and error mean squares. Using smaller blocks and row-column designs, measuring a related covariate, or using some form of nearest-neighbor analysis may control much of the extraneous variation in experiments.

Incomplete block designs are often useful in blocking for and controlling experimental variation. A very large number of tabled incomplete block designs are available. These may not meet all needs of experimenters, however. Incomplete block designs may be easily constructed for most situations encountered by experimenters. Two construction methods for use by experimenters are the ones presented in Patterson and Williams (1976) and Khare and Federer (1981). When the number of incomplete blocks equals the number of treatments, Federer (1993) presents another procedure for constructing incomplete block designs. The method is also useful for constructing row-column designs.

The shape of the eu may be altered in some cases to reduce variation within a block. Long narrow plots running perpendicular to variation gradients is one method for reducing variation among eu's within a block. Competition between treatments in adjacent eu's should be eliminated. This can be done by increasing space between eu's or by planting guard rows around eu's. The latter increases the size of an experiment. Assuming that competition does not exist may be very misleading. It is suspected that most field experiments reported in published literature that have small eu's have some element of inter-eu competition. Ignoring this fact may lead to incorrect conclusions from an experiment. Statistical analyses presented in statistical methods textbooks assume independence among eu's. The experimenter should verify whether the assumptions used for a statistical analysis are true or not. Using statistical analyses without considering their assumptions can lead to incorrect conclusions.

Treatment Design

The treatment design, selection of treatments to be included in an experiment, is an extremely crucial item for meeting the goals of an experiment. The exclusion of required treatments can lead to loss of information on certain goals of an experiment. Likewise, the inclusion of unnecessary treatments is a waste of space, material, and an experimenter's time. Controls or standards are necessary treatments for inclusion in an experiment to provide a point of reference for comparison. For intercropping experiments, the control may be a sole crop and/or the standard intercrop mixture for an area. For plant breeding experiments in intercropping systems, the choice of cultivar to intercrop with is important in evaluating breeding material (Callaway and Forcella, this volume, Chapter 7, discuss this in terms of crop-weed systems). In some maize breeding experiments, a single bean cultivar is often used as the other crop. If maize cultivars are to be grown with several bean cultivars in practice, a composite of the bean cultivars in the proportions they are typically grown with maize may be used to form the intercrop under which all maize lines are evaluated.

In general, any mixture of crops may qualify as a treatment for inclusion in an intercrop experiment. For certain goals and analyses, it is necessary to include sole crops as well as all possible combinations of mixtures. For several crops, several numbers of crops in a mixture, several densities of various crops, and other variables, the number of possible combinations becomes very large. Hence, an experimenter should choose the mixtures for inclusion in an experiment with great care. The goals of an experiment should be precisely defined, then treatments should be selected that will allow fulfillment of the goals. If a goal is to compare a group of mixtures with a standard mixture, these comparisons will only be possible if the standard mixture is included in the experiment.

In selecting a treatment design, the experimenter should follow these steps:

1. Precisely define goals
2. Select treatments allowing accomplishment of goals
3. Study the proposed statistical analyses
4. Decide in light of 1., 2., and 3. if the comparisons required are possible
5. Revise 1., 2., and/or 3. if 4. is not answered in the affirmative

Quite often it is possible to combine a number of proposed experiments and increase greatly the information obtained over separate experiments. Not only is more information usually obtained, but less experimental material and space may be required than for the separate experiments. For example, in a varietal test, an agronomist and an entomologist may want information on the same set of cultivars; rather than setting up two separate experiments, one experiment would suffice. As another example, instead of setting up two experiments to investigate various levels of two different factors, a factorial arrangement in one experiment could be performed.

Land Use and Agronomy

Probably the most used method for combining the results from an intercropping experiment is to use a form of land equivalent ratio (LER) (Willey and Osiru, 1972) or relative yield (deWit and van den Bergh, 1965). An LER is defined to be:

$$\text{LER} = \sum_{i=1}^{c} Y_{mi}/Y_{si} \tag{1}$$

where Y_{mi} equals yield of *ith* crop when in a mixture, Y_{si} equals yield of *ith* crop when grown as a sole crop, and c equals number of crops in a mixture. An LER as given above is poorly defined since it is not specified if Y_{mi} are individual plot yields or are means from r replicates of the *ith* crop from a given mixture. Likewise, Y_{si} could be obtained from a variety of sources, such as:

- Individual plot yields of sole crops.
- Mean yield from r replicates of sole crop i.
- Some theoretical "optimal" value for Y_{si}.
- Farmers' yields averaged over y years.
- Other.

Using all of the above possibilities, ten forms of LER could be obtained. There could be others; hence, it is absolutely necessary for an investigator to describe completely which LER has been selected for statistical analyses in order that others may comprehend the meaning of the results from an experiment.

If the denominators in the ratios of an LER are random variables, the statistical distribution of the LER statistics is unknown or not usable. For example, for normal random variates Y_{mi} and Y_{si}, the distribution of the ratio Y_{mi}/Y_{si} is a Cauchy distribution with infinite moments, that is, the parameters for the mean and variance of the variable Y_{mi}/Y_{si} are not defined (infinite) (Federer and Schwager, 1982). Since one ratio gives trouble, the sum of ratios, as in equation (1), is worse.

A normal variate ranges from minus infinity to plus infinity. Therefore, such characters as yield *cannot* be normally distributed since yields have a finite range starting at zero and bounded at the upper end. If yields are gamma or log normal distributed variates, then there is some hope for obtaining the statistical distribution of ratios and sums of ratios.

The following is one way out of the above dilemma. First, a base sole crop yield, say Y_{si}, is selected, then the LER from equation (1) is rewritten as an RLER:

$$\text{RLER} = \sum_{i=1}^{c} (Y_{sI}/Y_{si})Y_{mi} = \sum_{i=1}^{c} R_i Y_{mi} \tag{2}$$

where $R_i = Y_{sI}/Y_{si}$. Second, the ratios of yields R_i, are much more stable than are the $1/Y_{si}$ values; third, the R_i are taken as known constants. For example, from farmers' yields over many years in an area, it might be known that maize produces five times more kilograms per hectare (ha) than does bean. For Y_{sI} equal to maize sole crop yield and $R_b = Y_s$ maize$/Y_s$ bean$=5$, then for a maize-bean mixture RLER$=Y_m$ maize$+R_b Y_m$ bean$=Y_m$ maize$+5 Y_m$ bean.

239

If the denominators, Y_{si}, in (1) or the ratios R_i in (2) are constants, and if the Y_{mi} have a multivariate normal distribution, then the LER and the RLER have a normal distribution. Ratios of yields are more stable than actual yields; hence, using them as constants is more appropriate. It should be noted that the use of RLERs and of the R_i as constants affects the treatment design in that sole crop plots are no longer required. This is an important consequence, since reducing the size of the treatment design is an important item for most agricultural research projects.

The ratio Y_{mi}/Y_{si} indicates the relative yield of crop i in a mixture to crop i grown as a sole crop. If $Y_{mi}/Y_{si} = 1/2$ for all i, then there would be no advantage or disadvantage to using a mixture of crops over using sole crops. When none of the Y_{mi}/Y_{si} are less than 0.5 and some are larger, there would be an advantage to growing a mixture of crops. For some crop mixtures and specific cultivars, one of the important crops, say $i = 1$, may have $Y_{m1}/Y_{s1} = 1$. In such cases, any value of the other Y_{mi}/Y_{si} would result in an advantage for intercropping over sole cropping. In some cases, it has been observed that $Y_{m1}/Y_{s1} > 1$, which makes an intercrop even more advantageous. All of the above indicate that considerable care needs to be applied in selecting an intercrop mixture. The crops, the number of crops, the particular lines or cultivars of a crop, the density and intimacy of crops, and the relative importance of each element of the mixture are all considerations for selection of an intercrop mixture. The LER concept is further discussed in Willey (1979a,b).

Economic or Other Value

Various values may be assigned to the yield of crops in a mixture. For many people, value means monetary value. For other people, value is related to how well a crop satisfies the dietary goals of a family. For still others, value may be the frequency of produce for sale or barter throughout the year. Whatever value system is used, consider the value of a crop i to be P_i per unit such as a kilogram or an individual fruit. The value of crop i will then be $P_i Y_i$ where Y_i is the total yield of crop i per eu in kilograms, or number of fruit per eu. Since economic values may fluctuate considerably from year to year, we recommend the use of a ratio of price of crop i to a base crop, such as one, whenever all crop values are for the same unit, such as kilogram. In a mixture of crops, a linear combination of crop yields is the variable of interest to a grower. The following equations are similar to equations (1) and (2):

$$\text{Crop value } = V = \sum_{i=1}^{c} P_i Y_{mi} \tag{3}$$

where Y_{mi} is the yield of crop i in a c crop mixture and P_i is the price or value of a unit, such as a kilogram or individual fruit of crop i. For relative value and ratio of values to a base value, the equation is:

$$\text{RV} = \sum_{i=1}^{c} (P_i/P_1) Y_{mi} = \sum_{i=1}^{c} R_i Y_{mi} \tag{4}$$

RV values are recommended for use in summarizing information from an experiment.

Nutrition

In subsistence farming areas of the world, the number of calories provided by a crop is of vital importance. Protein content is also of importance for a proper diet. Therefore, in an intercrop experiment, it is necessary to assess the total calorie and/or protein content of a mixture in order to determine if it is more advantageous for a farmer to grow sole crops or intercrops. For example, suppose a farmer who derives food from farming needs a diet composed of four grams of carbohydrates to one of protein; suppose the maize cultivar grown produces nine grams of carbohydrate to one of protein and the bean cultivar produces one gram of carbohydrate to one gram of protein; suppose maize produces five times more kilograms per ha than does bean. If the farmer grows one ha of maize to eight ha of bean, the total produce will have approximately four grams of carbohydrates to one gram of protein. If the total carbohydrate requirement cannot be satisfied because the land is not available, the farmer may need to grow more maize and have a protein-deficient diet. He or she could also grow a mixture of maize and bean in combination with one of the sole crops.

A linear programming approach (Federer, 1993, Chapter 9, and discussion by B.R. Trenbath in Mead and Riley, 1981) may be used when the total carbohydrate and protein requirements are stated. The optimum number of hectares of each sole crop required to attain the goal is obtained. Likewise, a

241

linear programming approach may be used to determine the optimum distribution of acreage to achieve the stated goal using a combination of a sole crop and an intercrop.

For comparative purposes, the calorie conversion factor for a particular cultivar is known or can be obtained. Then the yield of crops in a c crop mixture may be converted to calories as follows:

$$C = \sum_{i=1}^{c} C_i Y_{mi} \tag{5}$$

where C_i is a calorie conversion factor for crop i. The same formula may be used to obtain total protein for the mixture. As before, a relative total calorie (or protein) may be obtained as

$$RC = \sum_{i=1}^{c} C_i Y_{mi}/C_i = \sum_{i=1}^{c} R_i Y_{mi} \tag{6}$$

where crop one was selected as the base crop and $R_i = c_i/c_1$. The only reason for using RC instead of C is for presentation purposes along with RLER and RV. The same graph may be used for all relative measurements. For interpretation purposes, formula (5) would be used.

Note that the $C_i Y_{mi}$ in equations (5) and (6) could be of a complex form in that carbohydrates, protein, fiber, and so on may be obtained for each crop in a mixture. The relative worth of protein to carbohydrate, fiber to carbohydrate, and so on may be available. A particular nutritional value for crop i would be

$$C_i Y_{mi} = (C_{ci} + R_{p/c} C_{pi} + R_{f/c} C_{fi} + \dots) Y_{mi}$$

where C_{ci}, C_{pi}, C_{fi}, and so on are the conversion factors for carbohydrate, protein, fiber, and so on for crop i; and $R_{p/c}$, $R_{f/c}$, and so on are the relative worths or values for protein, fiber, and so on, to carbohydrate. The above would combine all nutritional measurements into a single number for each crop in a mixture.

Multivariate Analysis

It has sometimes been recommended that the crops of a mixture of c crops be used as variates in a multivariate analysis and that a discriminant function analysis be used. As Federer and Murty (1987) demonstrate, this procedure leads to no usable information for an intercrop experimenter and is too restrictive. The mathematical criterion used to obtain a canonical variate (linear combination of yields) is to select the $\alpha\alpha_i$ in

$$\text{first canonical variate } = \sum_{i=1}^{c} \alpha_i Y_{mi} \qquad (7)$$

in such a way that no other selection of the α_i results in a larger ratio of treatment sum of squares divided by treatment plus error sums of squares for the first canonical variate. Then to the residuals, the above criterion is applied again to obtain a second canonical variate, such as

$$\text{second canonical variate } = \sum_{i=1}^{c} b_i Y_{mi} \qquad (8)$$

The procedure is continued until c canonical variates are obtained. As Federer and Murty (1987) point out, the α_i, b_i, and so on, have no practical interpretation and hence are not useful to the experimenter.

Previous sections in this chapter ("Land Use and Agronomy"; "Economic or Other Value") represent other types of multivariate analyses. Likewise, the multivariate procedure put forth by Pearce and Gilliver (1978, 1979) may be used to summarize information from an intercropping experiment. These multivariate methods are not limited to keeping the number of crops per mixture constant as is a discriminant analysis. The Pearce and Gilliver (1978, 1979) procedures are given for $c=2$ but could be extended to consider $c=3$ crops in a mixture. For LER, V, and C, the number of crops in a mixture may vary.

Soil Erosion and Structure

Sustaining yields in a farming system over long periods of time is highly dependent upon maintaining soil structure and reducing or eliminating soil erosion. Hence, in a breeding program it is important to evaluate lines for their performance relative to these characteristics and select an intercropping system that maintains soil structure and controls erosion even if sole crops would not. This demonstrates the complexity of evaluating farming systems to obtain sustainability.

One goal of an intercropping experiment may be to assess the amount of soil erosion and the changes in soil chemical, structural, and physical properties. For example, it is known that erosion in sole crop manioc may be high whereas erosion is drastically reduced when manioc is intercropped with melon, cowpea, or other crops (Aina et al., 1977; Lal, 1989). Likewise, in some intercrop mixtures on the same eu for two and more successive years, earthworm activity is significantly higher than in sole crop manioc (Ezumah and Hulugalle, 1989; Hulugalle and Ezumah, 1989). It would appear that certain intercrop mixtures may be beneficial in improving soil aeration, reducing soil bulk density, and building up soil organic matter (Lal, 1989).

The importance of intercropping to improving soil is related to soil structure, particularly in the major soils of humid tropics dominated by low-activity clays (Juo and Kang, 1987). These soils need additional organic matter to retain nutrients essential for plant growth. Reduced erosion from continuous vegetative cover provided through intercropping results in the retention of the topsoil and associated organic matter (Lal, 1989; Juo and Ezumah, 1990). Increase in soil infiltration, attributed to increased earthworm activity under intercropping, has been reported by Hulugalle and Ezumah (1989). Even in the highly fertile soils of temperate regions, mulching and/or maintenance of vegetative ground cover results in improved crop performance and soil conservation.

The amount of chemicals, soil aeration, and organic matter can be measured. But of what significance are they? Does their importance lie in explaining why and how yield over years is affected? Perhaps only yield should be measured and assessed. An unsolved problem here is how to use soil measurements (other than yield) more fully.

Continuous cropping of a mixture for many years will be required in order to measure changes in chemical content, changes in soil structure, and changes in organic matter. The various treatments will be compared with sole crop treatments and a standard intercrop treatment.

Density and Intimacy

Mead and Riley (1981) discuss the population density per ha for each of the crops in a mixture, intimacy (closeness of plants of crops to other crops), and spacing and arrangement factors. For sole crops, the problem is rather simple but becomes increasingly complex as the number of crops in a mixture increases. Many of these problems are discussed further in Federer (1993, Chapter 9). In addition, the ideas of parsimonious experiment design (PED), as described by Federer and Scully (1988), can be used effectively and efficiently to investigate optimal spacing, arrangement, intimacy, and density for each of the crops in a mixture. In formulating PEDs, use has been made of previous ideas like the Nelder fan and Okigbo-circle (Federer, 1993, gives references and description). Use of PEDs and proposed statistical analyses results in the desired information at considerable savings of time, material, labor, and money. Using PEDs involves changing ideas about setting up an eu and measuring the response from that unit. Functions of responses rather than single responses are used in the statistical analysis.

Biological Modeling

In developing biological theory for a system or procedure like intercropping, different statistical models and procedures from those presented in previous sections are required. It is not sufficient to simply compare cropping systems and mixtures. Knowledge of the biological processes governing why some systems or mixtures perform in the manner they do is necessary in order to develop methods for producing desired systems and mixtures in a more efficient manner. This situation has precedent in plant hybridization where diallel crossing, top-crossing, single-crossing, double-crossing, and multiple-crossing procedures and theory were developed and applied. Research in this area provides ample proof of the fact that it is not sufficient just to know that something happens, but it is necessary to know why.

Using some of the ideas from the above plus others, biological modeling for intercropping experiments was developed (Federer, 1993, Chapters 6 and 7; Federer, 1979; Federer and Raghavarao, 1987). For the models proposed, particular treatment designs are necessary. For example, all possible combinations of mixtures of c crops plus sole crops are necessary for some models such as those described by Federer and Raghavarao

245

(1987). For other models, a subset of the above treatment design suffices. Since the number of treatments can become large, it is necessary to use minimal designs that achieve the desired goal.

In setting up these models, effects such as general mixing ability (gma), bi-specific mixing ability, tri-specific mixing ability, and so on are discussed (Federer, 1979; Federer, 1993, Chapters 6 and 7). Gma refers to the ability of a crop or line to mix well with all others in the experiment. A cultivar with high gma should be included in mixtures. A cultivar with low gma does not mix well with any other cultivars in the experiment. When a pair of specific cultivars mixes well with each other but not with other cultivars, this is referred to as positive bi-specific mixing ability. When a particular triplet of cultivars is particularly good, we say it has high tri-specific mixing ability. The individual bi-specific effects of pairs may not be impressive but the combination of the three is. Statistical designs, models, and measures of all these effects are described in the above references.

In some situations when the items in a crop mixture are not identifiable, there will only be one response for a mixture from an eu. For example, an intermingled mixture of c lines of wheat with similar wheat kernels and plant type would be a case where the contribution from each of the lines was not available and only one response would be possible. If the c lines in a mixture were planted in an alternating plant fashion or if the c lines were in c adjacent rows, individual responses could be obtained. When the items in a mixture have quite different plant characteristics, such as maize and bean, the individual crop responses are readily obtained. Statistical models and analyses differ for the two cases, that is, only a single response for each eu or c responses, one for each crop in the mixture. Several models are available for each case (Federer, 1993).

Discussion

The comment that available statistical procedures are all that are needed to analyze data from intercropping experiments is shown to be incorrect. In analyzing and interpreting results from intercropping experiments, the level of thinking in going from sole crop data to two-crop mixture data goes up in difficulty by an order of magnitude. The amount of time and effort required for interpretation of results from an intercropping experiment is much greater than for a sole-cropping experiment. In analyzing data from an intercropping experiment, the following statement is appropriate: Expect the unexpected. Be prepared for a surprise over what sole-crop mentality would indicate. The complex biological mechanisms involved in the community living of plants may be quite different from when plants are living alone as a sole crop. If some of the results observed in past experiments hold up in general, biologists will need to advance their theories considerably in order to explain the results.

Statistical theory needs considerable extension to provide the necessary theory and methods for using such statistics as LER. It is surprising that so little statistical theory is available for ratios of random variables other than the binomial, Student's t, Snedecor's F, and the correlation coefficient. Scientists in many fields use ratios of random variables consistently, but statistical methods texts do not give the necessary theory and methods for dealing with ratios and sums of ratios of random variables. It would appear that statistical methodology should be extended to provide this theory. Performing a logarithmic transformation will not answer the problem for experimenters. As one sugar beet breeder said when told that a logarithmic transformation would make his data additive and eliminate interaction, "It is nonsensical to eliminate interaction because that is where all the profit is for the sugar beet growers and refineries! I sell interaction!" Transformation of yield responses from intercropping experiments when using such statistics as LER, RLER, V, VR, C, and RC would be invalid and uninterpretable. Uninterpretable measures are of no use to an experimenter, even though they might have nice statistical and mathematical properties. The statistics presented in the above sections on Land Use and Agronomy, Economic or Other Value, and Nutrition are interpretable and useful for experimenters even if the statistical properties for some of them are unknown.

References

Aina, P.O., R. Lal, and G.S. Taylor. 1977. Soil and crop management in relation to soil erosion in the rainforest of Western Nigeria. p.75–84. *In* G.R. Foster (ed.) Soil erosion prediction and control. Soil and Water

Conserv. Soc. Special Publication 21. Soil and Water Conserv. Soc., Ankeny, IA.

de Wit, C.T., and J.P. van den Bergh. 1965. Competition among herbage plants. Neth. J. Agric. Sci. 13:212–221.

Ezumah, H.C., and N.R. Hulugalle. 1989. Studies on cassava-based rotation systems in a tropical Alfisol. p.53. *In* Agronomy abstracts, ASA, Madison, WI.

Federer, W.T. 1979. Statistical designs and response models for mixtures of cultivars. Agron. J. 71:701–706.

Federer, W.T. 1984. Principles of statistical design with special reference to experiment and treatment design. p.77–104. *In* H.A. David and H.T. David (ed.) Statistics: An Appraisal. Iowa State University Press, Ames, IA.

Federer, W.T. 1993. Statistical design and analysis for intercropping experiments. Vol. 1. Two crops. Springer-Verlag, NY, NY.

Federer, W.T., and B.R. Murty. 1987. Uses, limitations, and requirements of multivariate analyses for intercropping experiments. p.269–283. *In* I.B. MacNiell and G.J. Umphrey (ed.) Biostatistics. D. Reidl, Dordrecht, The Netherlands.

Federer, W.T., and D. Raghavarao. 1987. Response models and minimal designs for mixtures of n of m items useful for intercropping and other investigations. Biometrika 74:571–577.

Federer, W.T., and S.J. Schwager. 1982. On the distribution of land equivalent ratios. Tech. Rept. Series No. BU-777-M. Biometrics Unit, Cornell University, Ithaca, NY.

Federer, W.T., and B.T. Scully. 1988. A parsimonious statistical design and breeding procedure for evaluating and selecting desirable characteristics over environments. Tech. Rept. Series No. BU-960-M. Biometrics Unit, Cornell University, Ithaca, NY.

Hulugalle, N.R., and H.C. Ezumah. 1989. Effect of cassava-based cropping system and rotation on soil physical properties of an Alfisol in Southwestern Nigeria. *In* Proc. 6th Int. Soil Conserv. Conf., Addis Ababa, Ethiopa, 6–18 Nov., 1989.

Juo, A.S.R., and H.C. Ezumah. 1990. Mixed root crop ecosystems in the wetter regions of Sub-Saharan Africa. *In* C.J. Pearson (ed.) Food crop ecosystems of the world. Elsevier, Amsterdam, Netherlands.

Juo, A.S.R., and B.T. Kang. 1987. Nutrient effects of modification of shifting cultivation in West Africa. p.450–470. *In* Proc. Int. Symp. on Nutrient Cycling in Tropical Forest and Savanna Ecosystems, Stirling, Scotland.

Khare, M., and W.T. Federer. 1981. A simple construction procedure for resolvable incomplete block designs for any number of treatments. Biometrical J. 23 (2):121–132.

Lal, R. 1989. Conservation tillage for sustainable agriculture: Tropics vs. temperate environments. Adv. Agron. 42:85–195.

Mead, R., and J. Riley. 1981. A review of statistical ideas relevant to intercropping research (with discussion). J. Royal Statistical Soc., Series A 144:462–509.

Patterson, H.D., and E.R. Williams. 1976. A new class of resolvable incomplete block designs. Biometrika 63(1):83–92.

Pearce, S.C., and B. Gilliver. 1978. The statistical analysis of data from intercropping experiments. J. Agric. Sci., Camb. 91:625–632.

Pearce, S.C., and B. Gilliver. 1979. Graphical assessment of intercropping methods. J. Agric. Sci., Camb. 91:625–632.

Willey, R.W. 1979a. Intercropping: Its importance and research needs. Part 1. Competition and yield advantages. Field Crop Abstracts 32:1–10.

Willey, R.W. 1979b. Intercropping: Its importance and research needs. Part 2. Agronomy and research approaches. Field Crop Abstracts 32:73–85.

Willey, R.W., and D.S.O. Osiru. 1972. Studies on mixtures of maize and beans (*Phaseolus vulgaris*) with particular reference to plant populations. J. Agric. Sci., Camb. 79:517–529.

APPENDIX

Organisms
Cited in the Text

PLANTS

alfalfa *Medicago sativa*

American grape *Vitis labrusca*

apple *Malus domestica*

Australian pine *Casuarina* spp.

avocado *Persea americana*

banana *Musa* spp.

barley *Hordeum vulgare*

bermudagrass *Cynodon dactylon*

bitter vetch *Vicia ervillia*

black wattle *Acacia mearnsii*

breadfruit *Artocarpus altilis*

broom millet *Panicum miliaceum*

bougainvillea *Bougainvillea* spp.

cabbage *Brassica oleracea* var.
  capitata

cashew *Anacardium occidentale*

coconut *Cocos nucifera*

chickpea *Cicer arietinum*

chrysanthemum *Chrysanthemum* spp.

coffee *Coffea arabica*

colonial bentgrass *Agrositis tenuis*

common bean *Phaseolus vulgaris*

common velvetgrass *Holcus lanatus*

cotton *Gossypium hirsutum*

cowpea *Vigna unguiculata*

crested dogtailgrass *Cynosurus*
  cristatus

cucumber *Cucumis sativus*

elephantgrass *Pennisetum purpureum*

European grape *Vitis vinifera*

foxtail millet *Setaria italica*

frangipani *Plumeria rubra*

giant foxtail *Setaria faberi*

green foxtail *Setaria viridis*

groundnut *Arachis hypogaea*

guaje rojo *Leucaena esculenta*

guineagrass *Panicum maximum*

hibiscus *Hibiscus rosa-sinensis*

horsetail-tree *Casurina equisetifolia*

jemara *Casuarina junghuhniana*

jute *Corchorus* spp.

keawe *Prosopis pallida*

lambsquarters *Chenopodium album*

lentil *Lens culinaris*

lettuce *Lactuca sativa*

leucaena *Leucaena leucocephala*

mahogany *Swietenia mahogani*

maize *Zea mays*

mangium *Acacia mangium*

mango *Mangifera indica*

manioc *Manihot esculenta*

melon *Cucumis melo*

monkeypod *Samanea saman*

Monterey pine *Pinus radiata*

mungbean *Vigna radiata*

neem *Azadirachta indica*

oat *Avena sativa*

oil palm *Elaeis guineensis*

onion *Allium cepa*

papaya *Carica papaya*

pea *Pisum* spp.

pearl millet *Pennisetum typhoides*

pepper *Capsicum* spp.

petunia *Petunia* spp.

pigeon pea *Cajanus cajan*

pine *Pinus* spp.

pineapple *Ananas comosus*

poorjoe *Diodia teres*

poplar *Populus* spp.

potato *Solanum tuberosum*

poverty oatgrass *Danthonia spicata*

purslane speedwell *Veronica peregrina*

quickstick *Gliricidia sepium*

rapeseed *Brassica napus*

red cedar *Cedrela* spp.

red leucaena *Leucaena diversifolia*

redroot pigweed *Amaranthus retroflexus*

rice *Oryza sativa*

river sheoak *Casuarina cunninghamiana*

rosewood *Dalbergia* spp.

rubber *Hevea brasiliensis*

ryegrass *Lolium perenne*

Scots pine *Pinus sylvestris*

sorghum *Sorghum bicolor*

soybean *Glycine max*

squash *Cucurbita pepo*

sugarcane *Saccharum* spp.

sweet corn *Zea mays*

sweet vernalgrass *Anthoxanthum odoratum*

tall fescue *Festuca arundinacea*

tea *Camellia sinensis*

teosinte (perennial) *Zea perennis*

timothy *Phleum pratense*

tobacco *Nicotiana tabacum*

tomato *Lycopersicon esculentum*

triticale *Triticosecale* spp.

wheat *Triticum aestivum*

white clover *Trifolium repens*

yam *Dioscorea* spp.

yellow nutsedge *Cyperus esculentus*

INSECTS

cowpea curculio *Chalcodermus aeneus*

cucumber beetle *Acalymna* spp.

European corn borer *Ostrinia nubilalis*

grape phylloxera *Phylloxera vitifolae*

green leafhopper *Nephotettix virescens*

Hessian fly *Phytophaga destructor*

leucaena psyllid *Heteropsylla cubana*

northern corn rootworm *Diabrotica longicornis*

potato aphid *Macrosiphum euphorbiae*

sorghum greenbug *Schizaphis graminum*

tobacco hornworm *Protoparce sexta*

tomato fruitworm *Helicoverpa zea*

western corn rootworm *Diabrotica virgifera*

DISEASES

aflatoxin *Aspergillus flavus* and *A. parasiticus*

apple spot *Pseudomonas papulosum*

brown leaf spot of coffee *Cercospora coffeicola*

gray leaf spot *Cercospora zeae-maydis*

late blight *Phytophthora infestans*

phytophthora root rot *Phytophthora megasperma* var. *sojae*

pink root of onion *Pyrenochaeta terrestris*

powdery mildew of wheat *Erysiphe graminis*

rice blast fungus *Pyricularia oryzae*

snowmolds *Typhula* spp. and *Fusarium nivalea*

southern corn leaf blight *Bipolaris maydis*

southern rust *Puccinia polysora*

wheat rust *Puccinia* spp.

· · ·

The Contributors

David M. Bates is a professor of biology and society at the Liberty Hyde Bailey Hortorium of Cornell University. His numerous publications on ethnobotany and plant systematics include *Biology and Utilization of the Cucurbitaceae* (coedited with R.W. Robinson). He has been actively involved in a number of scientific societies, serving as president of the Association of Systematics Collections, the American Society of Plant Taxonomists, and the Society of Economic Botany.

James L. Brewbaker is a professor of horticulture at the University of Hawaii. He has authored two-hundred scientific publications, including his textbook *Agricultural Genetics,* which has been translated into seven languages.

M. Brett Callaway is tropical germplasm manager in the Department of Corn Breeding for Pioneer Hi-Bred International, Inc. He has published scientific articles on a wide range of topics, including weed ecology, new crop development, and plant breeding.

W. Ronnie Coffman is associate dean for research, College of Agriculture and Life Sciences, and director of Agriculture Experiment Stations at Cornell University. Dr. Coffman has been instrumental in the development of improved rice varieties now grown on several million hectares worldwide.

He has served as a member of the board of trustees of the West Africa Rice Development Association and of Winrock International.

Donald N. Duvick is retired senior vice-president of research for Pioneer Hi-Bred International, Inc. and an affiliate professor of agronomy at Iowa State University. He has published a number of influential research articles on maize and has been actively involved in several professional organizations, including serving as president of the Crop Science Society of America and the American Society of Agronomy.

Walter T. Federer is Liberty Hyde Bailey Professor Emeritus in the Biometrics Unit of the Department of Plant Breeding and Biometry at Cornell University. He has authored a number of popular textbooks, including *Experimental Design, Statistics and Society,* and *Statistical Design and Analysis for Intercropping Experiments: Two Crops.*

Frank Forcella is a research agronomist for the USDA-ARS and adjunct assistant professor of agronomy at the University of Minnesota. Dr. Forcella has over fifty scientific publications relating to ecology and weed science.

Charles A. Francis is a professor of agronomy and extension crops specialist at the University of Nebraska. Dr. Francis has published extensively on sustainable agriculture and related topics. Books which he has edited include *Multiple Cropping Systems* and *Sustainable Agriculture in Temperate Zones* (with C.B. Flora and L.D. King).

Major M. Goodman is Reynolds Distinguished University Professor in the Department of Crop Science at North Carolina State University. He has published numerous scientific articles and book chapters on maize germplasm. His honors include election into the United States National Academy of Sciences.

Michael P. Hoffmann is an assistant professor of entomology at Cornell University. He has published a number of scientific articles on agricultural entomology.

Molly Kyle is an assistant professor of plant breeding and biometry at Cornell University. Her accomplishments include editing the book *Resistance to Viral Diseases of Vegetables: Genetics and Breeding*.

Susan McCouch is an associate geneticist for the International Rice Research Institute and adjunct assistant professor of plant breeding and biometry at Cornell University. She has published several scientific papers on the molecular genetics of rice and was awarded the Richard Bradfield Award for significant international contributions to crop protection and production.

Jim R. Myers is an assistant professor of plant breeding and genetics at the University of Idaho. Dr. Myers has published a number of scientific articles on topics relating to plant breeding.

Pamela C. Ronald is an assistant professor in the Department of Plant Pathology at the University of California at Davis. She has published scientific papers on the molecular genetics of plant resistance to bacterial pathogens.

Brian T. Scully is an assistant professor of vegetable crops at the University of Florida Institute of Food and Agricultural Science. He has published over twenty-five scientific articles on quantitative genetics and plant breeding.

Margaret E. Smith is an assistant professor of plant breeding and biometry at Cornell University. Dr. Smith worked at the International Maize and Wheat Improvement Center in Mexico and the Tropical Agricultural Center for Teaching and Research in Costa Rica before joining the faculty at Cornell.

Edward J. Souza is an assistant professor of plant breeding and genetics at the University of Idaho. Dr. Souza has published over twenty scientific papers on quantitative genetics and the improvement of small grains.

H. David Thurston is a professor of plant pathology at Cornell University. He is internationally recognized for his work on plant pathology in the developing nations of the Tropics. His books include *Sustainable Practices for Plant Disease Management in Traditional Farming Systems*.

Index

Agriculture: energy inputs, 23; origins of, 20, 21; outputs, 23
Allomones, 85
Antibiosis, 83

Bio-pesticide, 168
Biotechnology, concerns relating to use of, 176
Breeding, 151–153; objectives of, 61–64; principles of, 151–153; strategies of, 65–67; trees and shrubs, 135–136, 146–147

Cell culture, uses of, 163–165
CIMMYT (International Maize and Wheat Improvement Center), 217
Competitive ability: breeding for, 109; important traits, 115; in natural systems, 105; in pastures, 105; in soybean, 106–115; with weeds, 105
Cucurbitacin, 85

Diseases: crop resistance to, 48, 62, 81, 139–141; crop tolerance to, 48, 139–141

Diversity, genetic, 55
DNA markers, 160–161; randomly applied polymorphic DNA (RAPD), 160–161, 162; restriction fragment length polymorphism (RFLP), 160–161, 176

Ecovalence, 212, 221, 224
Effective gene pool, 136
Environment classification, 214
Experimental unit, 234; definition of, 235; shape of, 236
Experiment design, 234, 235–237

Food supply, 23
Forest, loss of, 133

Gene transfer, 170–171
Genetic diversity, reduction of, 159
Genetic resources, forest, 134
Genetic variation, assessing, 161–162
Genotype by Environment Interaction (G x E) AMMI analysis, 206–207, 215, 216, 218, 224; analysis of, 196–197; classification of models, 197–199; cluster analysis, 205–206; correlation

259

Other volumes in the series Our Sustainable Future:

Volume 1
Ogallala: Water for a Dry Land
John Opie

Volume 2
Building Soils for Better Crops: Organic Matter Management
Fred Magdoff

Volume 3
Pesticide-Free Farming in the Midwest
Jim Bender